INSIDE BUBBLE EARTH: CLIMATE CHANGE

DES HUNT

OneTree HOUSE

First published by OneTree House Ltd,
New Zealand

© Des Hunt, 2021

All rights reserved. No part of this publication may be reproduced, stored in a retrieval system or transmitted in any form or by any means, electronic, mechanical, photocopying, recording or otherwise, without the prior permission of the publisher.

Design: Vasanti Unka

Printed: YourBooks, Wellington, New Zealand

9 8 7 6 5 4 3 2 1   1 2 3 4 5/2

Supported by:

PHOTO CREDITS:

The author acknowledges the work of the following photographers listed in order of their appearance:

Shutterstock: (8) Lennart Nilsson, (9) Solodov Aleksei, (14) Esteban De Armas, (16) Jose Luis Calvo, (19) Lovelyday 12, (19) Showcake, (20) Lana Mais, (20) Dejan Stanic Micko, (22) Nicholas Primola, (23) Sklyarov Roman, (23) Igor Barin, (33) Damsea, (34) Han Engbers, (35) Tarcisio Schnaider, (40) Creativa Images, (40) fivepointsix, (43) Shpatak, (43) Solarisys, (43) Sabangvideo.

(41) © Stuff NZ.

CONTENTS

INTRODUCTION.............4
Shapes in the Dark........................4
Shining the Light..........................6

1 BEGINNINGS 8
Bubbles Big and Small 8
Climate.......................................9
A Climate Survival Story................10
Variation....................................13

2 SCIENCE OF CLIMATE CHANGE 15
Living Things 15
Building Blocks 16
Plants 19
Animals 20
Carbon 21
Light...22
Properties of Light.......................24
Greenhouse Effect 26

3 CAUSES AND CURES......28
Fossil Fuels and CO_2 28
Returning the Balance 30
Carbon Footprints 31
Agriculture and Methane 32
Zero Carbon 34
Renewable Energy.......................35
Hydrogen and Fuel Cells 37

4 CONSEQUENCES...........39
Extreme Weather Events 40
Ocean Acidification 42
Mass Extinctions 44

5 IN THE END 46
Personal Responsibility 46
Acitivity: Grow Plants 48
Glossary....................................50
Further Reading 54

In this book you may find words you are not familiar with. Try looking at the glossary on page 50 to see if the word is explained there.

INTRODUCTION

Shapes in the Dark

At first Alex had enjoyed the family day out at the regional park. The picnic featured all their favourite food and drinks. The games they played afterwards were exciting but exhausting. Only later in the day had things gone wrong – when Dad suggested they collect pine cones to take home as firelighters. Each person was given a sack with the promise of a special treat for whoever collected the most.

By then the sun had set and the temperature was dropping quickly. But Alex wasn't worried and ran off to a place where they'd spotted lots of pine cones earlier. The others all went in different directions, which suited Alex fine: there would be no need to share the special treat.

Alex had almost filled the bag before realising how gloomy it had become. The only sounds were those from the darkened forest – rustling and creaking and groaning. Then Alex sensed, more than saw, a creature nearby. A pair of glowing eyes, a blood-red open mouth, surrounded by the dark shape of a fearsome body. And panting sounds as if the creature had run to get to this spot especially to attack a child.

Alex screamed, dropped the bag and ran, giving no thought to direction, other than to get away from the terrible beast.

Fortunately, Alex hadn't wandered too far from the camping area, and soon others were there, hugging and comforting.

When asked what had been so scary, Alex replied there was something in the forest, a shape with glowing eyes. Dad reckoned it could have been one of the animals that were farmed in the park, its eyes reflecting the lights from the toilet block. But Alex knew otherwise. There was a monster in the dark forest, one that survived by attacking children.

That monster featured in Alex's nightmares for many years, always with glowing eyes and a bloody mouth. Because of that, the family never returned to the regional park for a picnic and games, or to pick up pine cones. They kept away from the place just in case Alex had been right.

Shining the Light

Fear of the unknown is common to all of us, child or adult. Often it is associated with darkness when we are unable to see and understand what is happening. Monsters under beds, or things that go bump in the night, are common starting points for horror stories. If there had been more light in the forest where Alex was gathering pine cones, it would have been obvious what the monster really was – a lamb, frightened because it had lost its mother. In that case, Alex's reaction would have been one of concern for the lamb, rather than fear of a monster.

Global climate change is like a monster. Most of the time it stays hidden and we can live our lives without knowing it exists. Then a storm strikes, worse than any seen before, or a forest fire kills more animals and people than any other fire has. That's when we know the monster is out there and we become worried about the future of our planet.

The purpose of this book is to shine a light on climate change so we clearly see what it is and how it might be tamed. Once we understand the causes and effects, we will see ways to keep it under control and, hopefully, reverse some of the damage already done.

Humankind has dealt with global-sized problems before, the most recent being a virus called COVID-19 – a scary, killer disease that spread around the world in 2020. We dealt with that by forming bubbles: country-sized bubbles, city bubbles, and – the one most of us will remember – the family bubble. We learnt to wash our hands, wear face masks, keep our distance from others. In other words, we followed instructions and did the right thing – we cooperated, working together as one. While this didn't destroy the COVID monster, it kept it under control until vaccines could give us immunity.

The bubble method worked because the COVID virus needs humans to carry it around.

But climate change happens because of changes in the atmosphere, and there is no way we can isolate parts of the atmosphere into smaller bits. There is only one bubble and that is the *whole planet*. But the same ideas apply. If we understand what is happening and all work together, we can deal with climate change and, in the process, create a better Bubble Earth for everyone.

1 BEGINNINGS

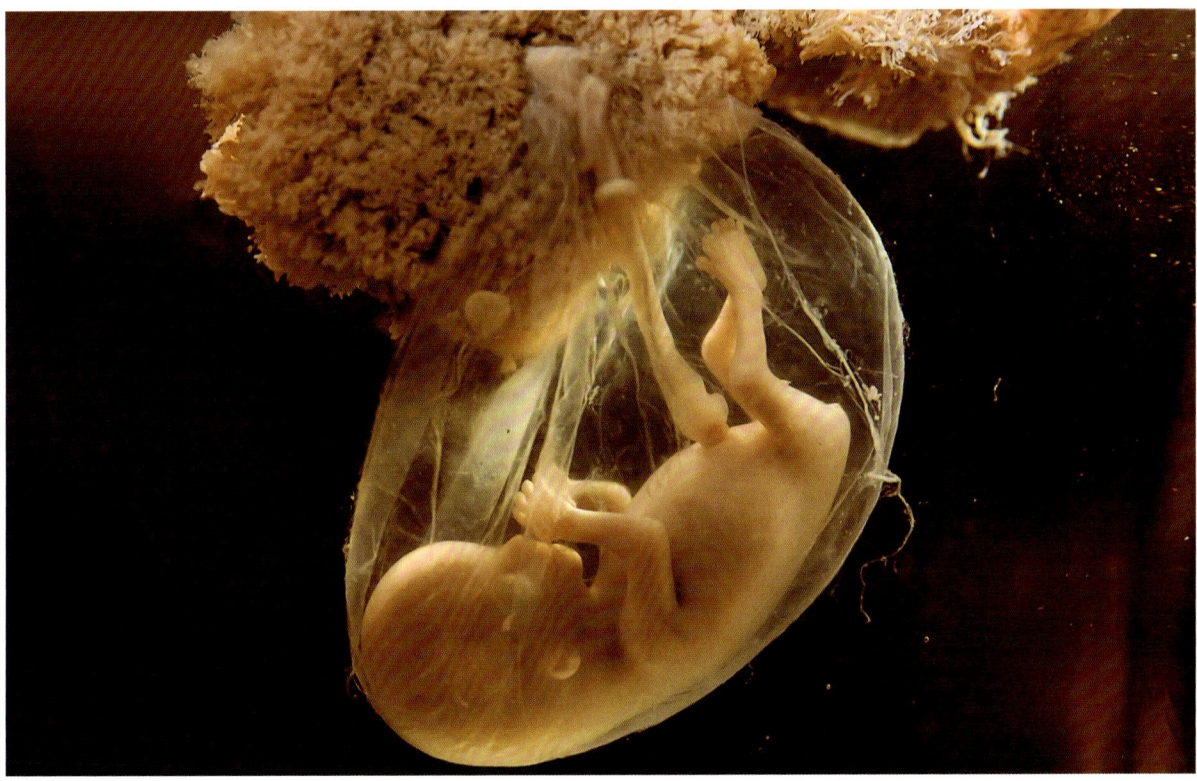

Bubbles Big and Small

We all begin life in a bubble. Every living thing does. The baby human shown above will spend nine months in its liquid bubble before moving into a bigger space. During those months, it is totally dependent on the mother for all its needs, especially water, oxygen and food.

Simpler animals have simpler bubbles. The squid (top, right) floats around the ocean in its bubble for weeks before being set free. When the egg was formed, it was given enough food and fresh water for the baby squid to survive and develop. All it needed from outside was oxygen.

Birds' eggs are even better stocked. They have a supply of oxygen as well as water and food. The air cell of a chicken egg can be seen in the opened egg pictured (right). Next time you eat a boiled egg, see if you can find that air cell.

Even though eggs have a supply of air, there's not enough to last until the chick hatches. Towards the end of development, oxygen must pass in through tiny holes in

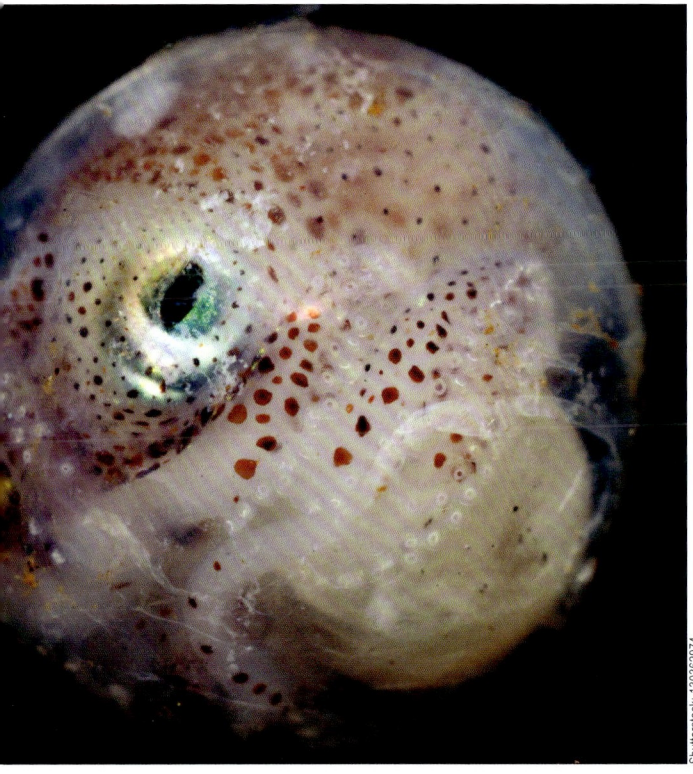

the shell, and carbon dioxide must pass out. All living bubbles must exchange something with the outside to survive.

As we will find, Bubble Earth is no different. It is totally dependent on something coming in from space, and something going out into space. It is the balance between the ins and outs that sets the climate at the surface. But before we look at how this works, we need to define climate.

Climate

Climate is the word we use to cover weather over a long time, like several years. The term includes all the seasons and the differences between one year and another. We say that places like the tropics have a hot climate, even though it may sometimes be frosty at night in winter. The poles have a very cold climate, although summer days in Antarctica can often be quite mild in places. Climate is an idea that averages things out.

Without doubt, the climate inside Bubble Earth is changing. It has changed in the past and will change in the future. In this first section, we'll look at changes in the past through the eyes of a species that has been around a lot longer than we have. Hopefully, it will help us to understand the climate change that is happening right now.

A Climate Survival Story

The image left shows a baby tuatara struggling to get free of her egg bubble where she's been developing for the last 12 months. Her name is Tahi and she is the first of the eight eggs in the nest to hatch. Not that she knows it, for she's too busy climbing out of the egg and dirt to worry about anything other than herself.

Her mother laid and buried the eggs in October 2020. This day – her hatch-day – is 15 October 2021. If she's anything like her father, she'll celebrate many, many hatch-days, perhaps live to over a hundred. With a bit of luck, she'll see the year 2121. Certainly her children could, and her grandchildren could even see 2222. But, then again, they might not. It all depends on climate change.

Tuatara have seen lots of climate changes. Tahi's ancestors have seen very hot times and very cold times. The hottest was 20 million years ago when the average temperature was five degrees higher than today. The climate in New Zealand back then was tropical. Fossils of crocodiles have been found in the South Island next to the bones of a tuatara. There was also a giant parrot, almost a metre tall, and even a small mammal. The crocodile, mammal and parrot became extinct, but tuatara lived on, as they have done for 250 million years.

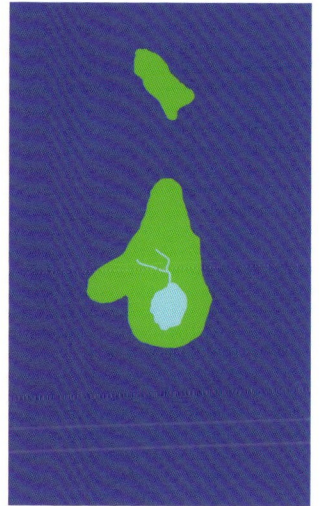

AOTEAROA NEW ZEALAND
As it may have been 20 million years ago.

During the last Ice Age, 20,000 years ago, it was bigger than it is today.

New Zealand wasn't as big 20 million years ago, mostly because the main mountain range had not yet formed, but also because the sea level was 20 metres higher than at present. Changing sea level happens with climate change. As the temperature goes up, so does the sea level. It also works the other way round.

The coldest time Tahi's ancestors experienced was 20,000 years ago, during the Ice Age. At that time, a third of the South Island was covered with ice, summer and winter. Because so much water was trapped in ice around the world, the sea level was 100 metres below what it is today. This meant New Zealand was one large island, as shown by the map. The green-coloured areas to the west (left of present-day New Zealand) were covered in forest; the brown to the east was mostly grassland and tussock. Tuatara may have lived all over the land, although it would have been difficult in the south where temperatures were cold most of the year.

Temperature controls the development of tuatara eggs in two main ways: (1) the time it takes for them to incubate and (2) whether the baby that hatches is male or female.

Tuatara eggs are laid in burrows in the ground. There is no heating apart from sunshine falling on the earth. The colder the climate, the longer the eggs take to develop. The opposite is also true: as the temperature continues to increase over this century, the time for the eggs to hatch will shorten. This may be a good thing, as it reduces the time predators such as rats will have to dig up nests and eat the eggs. But if it means the eggs hatch in autumn or winter, when there is less food, it will not benefit tuatara at all.

The second effect of temperature rise certainly won't help the animal in any way. This is the setting of the sex of individuals. In humans, the sex of the baby is controlled by the DNA of the parents. With tuatara, it is decided while the eggs are incubating.

The important time is three months after the eggs are laid. As most are laid in spring, this makes January and February the time when the sex is set as either male or female – that's the hottest time of summer. If the temperature at that time averages between 20 °C and 22 °C then the tuatara are likely to be a mix of males and females. In the diagram, this is shown by sex symbols on the eggs – ♂ for males, ♀ for females. If the temperature is below 20.5 °C, then all the babies that hatch will probably be female. But if the temperature is above 22.5 °C, they'll all be male.

If climate change in New Zealand keeps going as it is, the temperature will increase one degree by 2040 and two degrees by 2090. This means that all the eggs hatching at the end of this century would be male, which would be disastrous. No females means no eggs, no hatching tuatara, and no future for the species.

That's why Tahi's children and grandchildren may run into serious trouble later this century. If all eggs hatch as males then, without human help, tuatara will be doomed and in danger of extinction.

"But ... but ... If this is true," you might ask, "how did they survive the Ice Age? Wouldn't the cold climate produce only female tuatara? Isn't 'all females' just as bad as 'all males'? And wouldn't it also have been a problem during the hot time 20 million years ago? Is there more to this than just temperature?"

The answer is yes, and the reason is that, like humans, not all tuatara are the same.

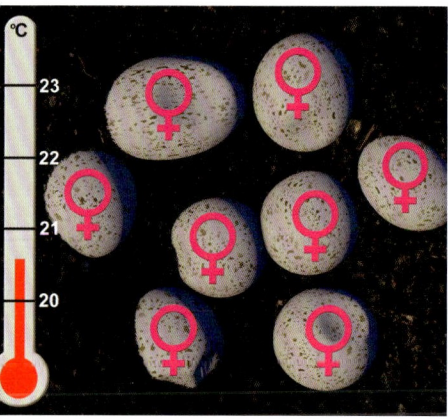 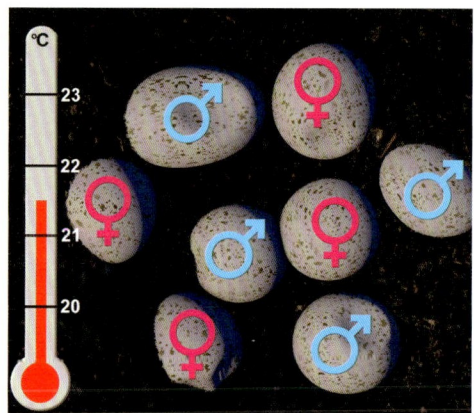 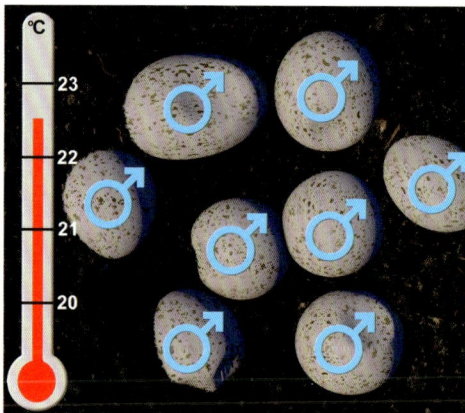

Variation

Every species changes with time, including humans. You are different from your parents, and your children will be different from you. This is called variation. Over long periods of time, variation allows a species to cope with changing conditions.

The average time between hatching and becoming a parent for tuatara is about 20 years. This is called the generation period.

To find the number of generations since the Ice Age we divide 20,000 years by 20 years giving 1000 generations. This is a lot of chances for things to change. The tuatara of today probably look much the same as those which lived during the Ice Age, but the chemistry in their bodies and eggs will be different.

The temperature increase since the Ice Age is five degrees. That's one degree rise every 4000 years. Adjusting to such slow change is easy. But with the climate expected to warm up by at least two degrees this century, there is simply not enough time for the tuatara to adjust.

However, there is hope. The last time there was a rapid change of climate on this scale was 66 million years ago, when Earth

No two seeds on this corn cob are the same even though they have been formed by the same parent. The difference in shape and size comes from how they grew on the cob; the difference in colour is due to variation in their DNA.

was struck by an asteroid. This filled the atmosphere with soot and other small particles, blocking the sunlight and creating a much colder climate over the entire globe. Although this lasted for less than a century, it was long enough for lots of species to become extinct, including most of the dinosaurs – only the bird-like ones survived.

Tuatara also survived, probably because the cold time lasted less than their normal lifetime. They couldn't breed in the cold, but enough of them survived to continue the species when the climate returned to normal.

If the climate change of our century can be slowed or, better still, reversed, there will be less of an impact on all species. To do this, we need to understand the science behind climate change. Only then can we look at what action is needed. If we get this right then Tahi will end up having great-great-*great*-grandchildren, and she might even still be around to see them.

2 SCIENCE OF CLIMATE CHANGE

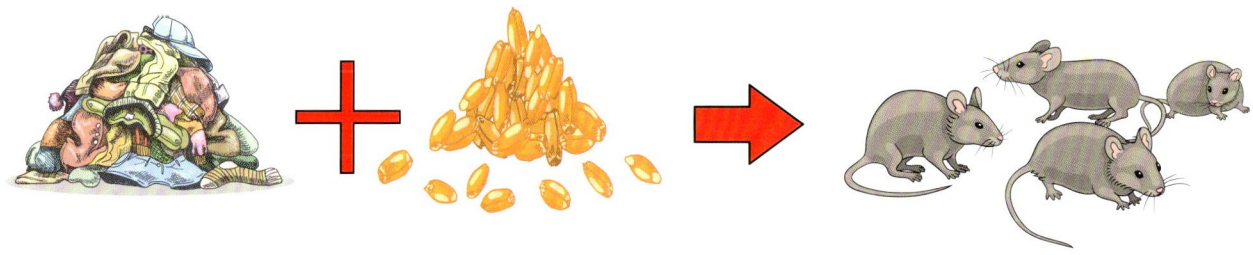

Living Things

You probably have a good idea of what is living and what isn't. And I'm sure you don't expect something that is not living, like your stinky clothes, to turn into something that is living, such as a mouse. But this was not always the way people thought. Back in the 1600s, people had some strange ideas about how living things formed.

One idea that got publicity was that mice could be made by combining sweaty clothes with wheat. If you left them together for 21 days, you ended up with mice. Another favourite was horse hairs on the ground turning into worms when it rains.

These ideas are nonsense, of course, and can be easily proved wrong by simple experiments. Science is about testing ideas and modifying our thinking, based on the results. One of the first people to perform scientific experiments was Joseph Priestley in the 1700s. One of these is important when discussing climate change. He was investigating air and its importance to living things.

In the first part of the experiment, he put a mouse and a burning candle in a sealed glass jar, creating a bubble. He found that both the flame and the mouse died.

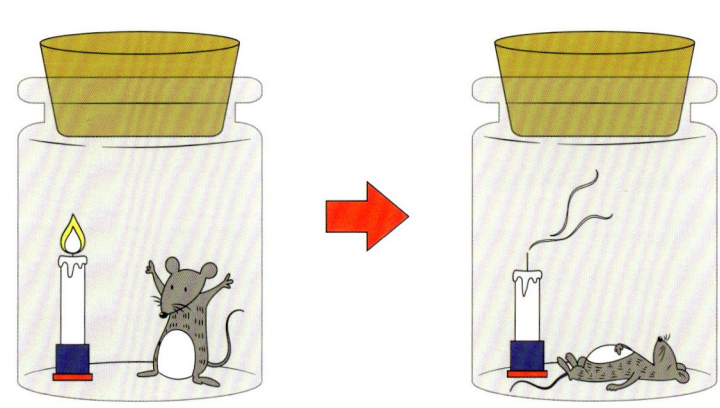

Then he repeated it with a plant in the bubble. Everything lived, at least for a while. This was the first scientific evidence that animals need plants to survive. The connection between plants and animals is the basis of all understanding of climate change.

In science, an animal is any living thing that must eat food to survive. In everyday life, we tend to think of animals as fish, amphibians, reptiles, birds or mammals – the ones that have bones. But insects, worms and slugs are also animals. The diagram on the facing page shows the range of different types of animals.

Humans are also animals as far as science is concerned. Even though calling someone an animal is usually thought of as rude, it's okay if you're talking science.

A plant is any living thing that can make its own food. Most of them growing on land are green. Some that grow in the sea may be red or brown. The colour of a plant is important in the process that makes food. So that we can understand how this works, we need to look at the building blocks of matter.

Building Blocks

All living things are made from building blocks, although the pieces are nowhere near as big as the Lego blocks shown. Living blocks are called cells and in something as big as a monkey, there are about four trillion of them. That's four followed by 12 noughts.

Even though cells are tiny, they are made from building blocks too. We call these smaller blocks – atoms.

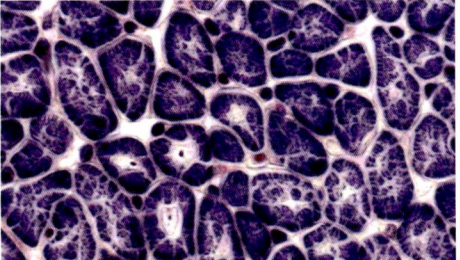

Everything you can touch is made of atoms.

16

Types of Animals

Types of Plants

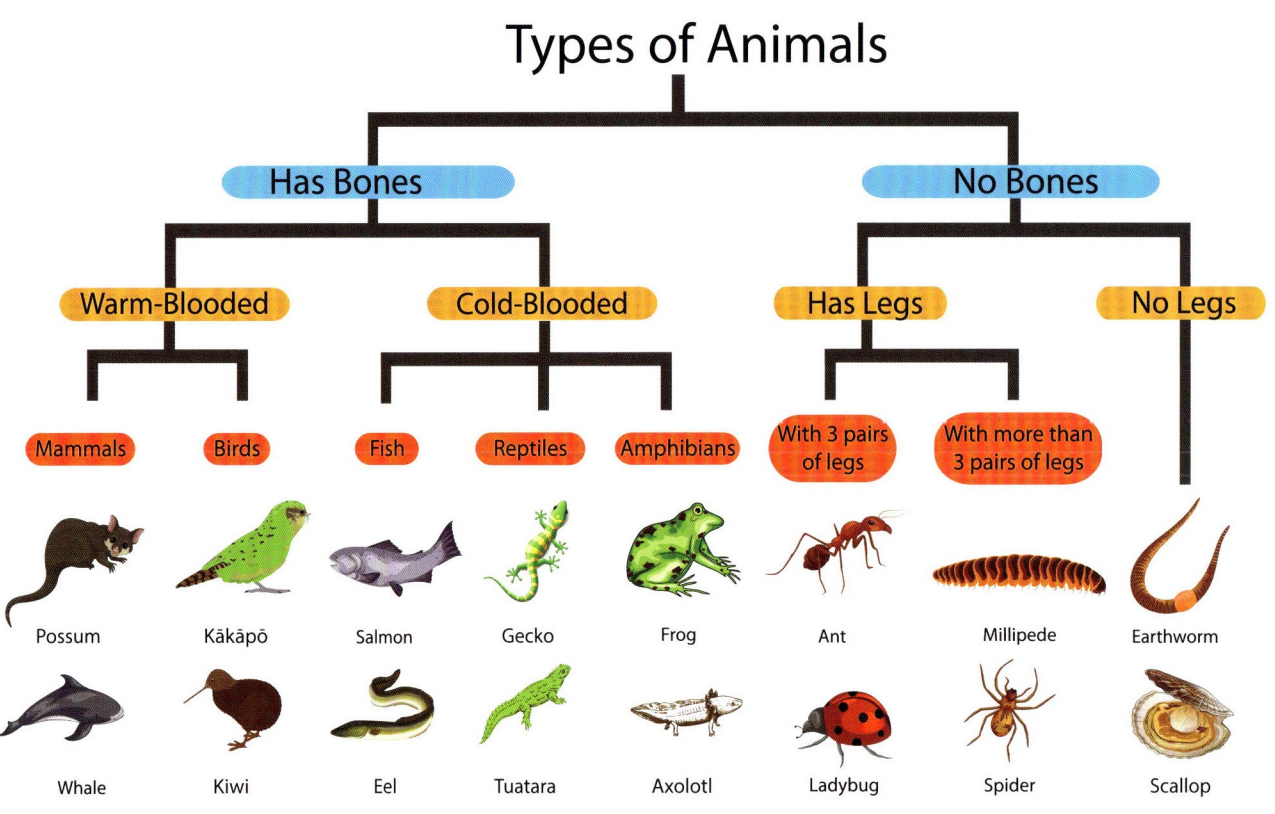

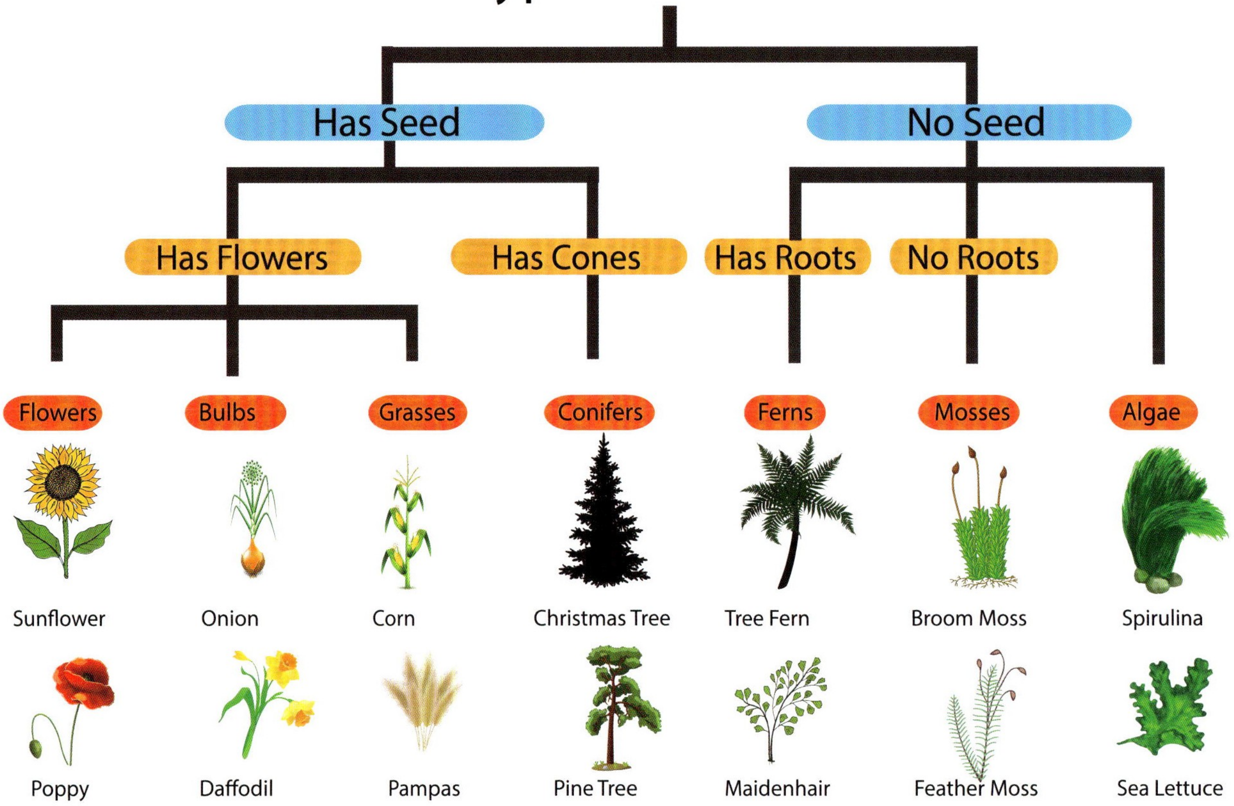

Usually atoms are combined together to form molecules. You've probably seen the formula H_2O used for water. That says that a molecule of water has an atom of oxygen and two of hydrogen. We could use Lego blocks to model molecules, but we'll use the normal scientific construction set which has atoms as spheres.

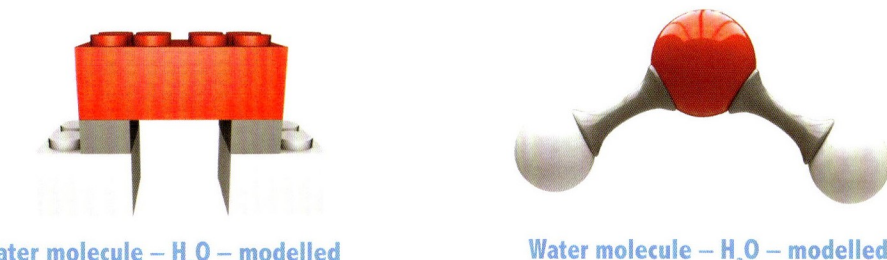

Water molecule – H_2O – modelled using Lego

Water molecule – H_2O – modelled using a molecular construction set

In this study of climate change we need only four types of atoms:

Oxygen   Hydrogen   Carbon   Nitrogen

All the substances involved in creating climate are made from combinations of these four atoms. The main molecules with their formulas are shown below.

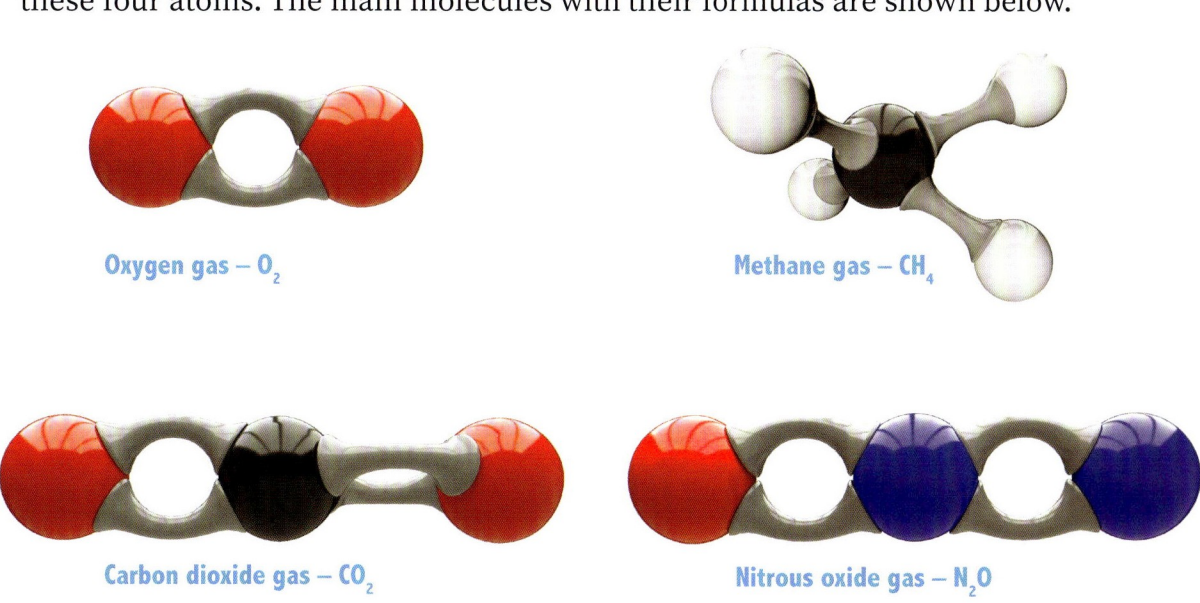

Oxygen gas – O_2

Methane gas – CH_4

Carbon dioxide gas – CO_2

Nitrous oxide gas – N_2O

Plants

Plants get their energy from the sun.

They make their own food using carbon dioxide from the air, and water from the soil. This process is called photosynthesis.

The molecules in all foods are combinations of lots of carbon, oxygen and hydrogen atoms. Glucose is the simplest of these molecules.

Any food the plant doesn't need to grow and survive is stored in roots, stems, trunks and fruit.

Photosynthesis also produces oxygen which is released into the air.

Photosynthesis

$$CO_2 + H_2O \rightarrow GLUCOSE + O_2$$

Six carbon atoms from the atmosphere are stored as food

Animals

Animals get their energy from eating food.

They burn food with oxygen from the air in a process called respiration.

The food comes either from a plant or from an animal that has eaten a plant.

All food molecules contain carbon, oxygen and hydrogen atoms. Some, like proteins, also contain nitrogen atoms.

Any food that isn't needed to grow and survive is stored as fat.

Respiration produces carbon dioxide and water which are released into the environment. Respiration is the reverse reaction to photosynthesis.

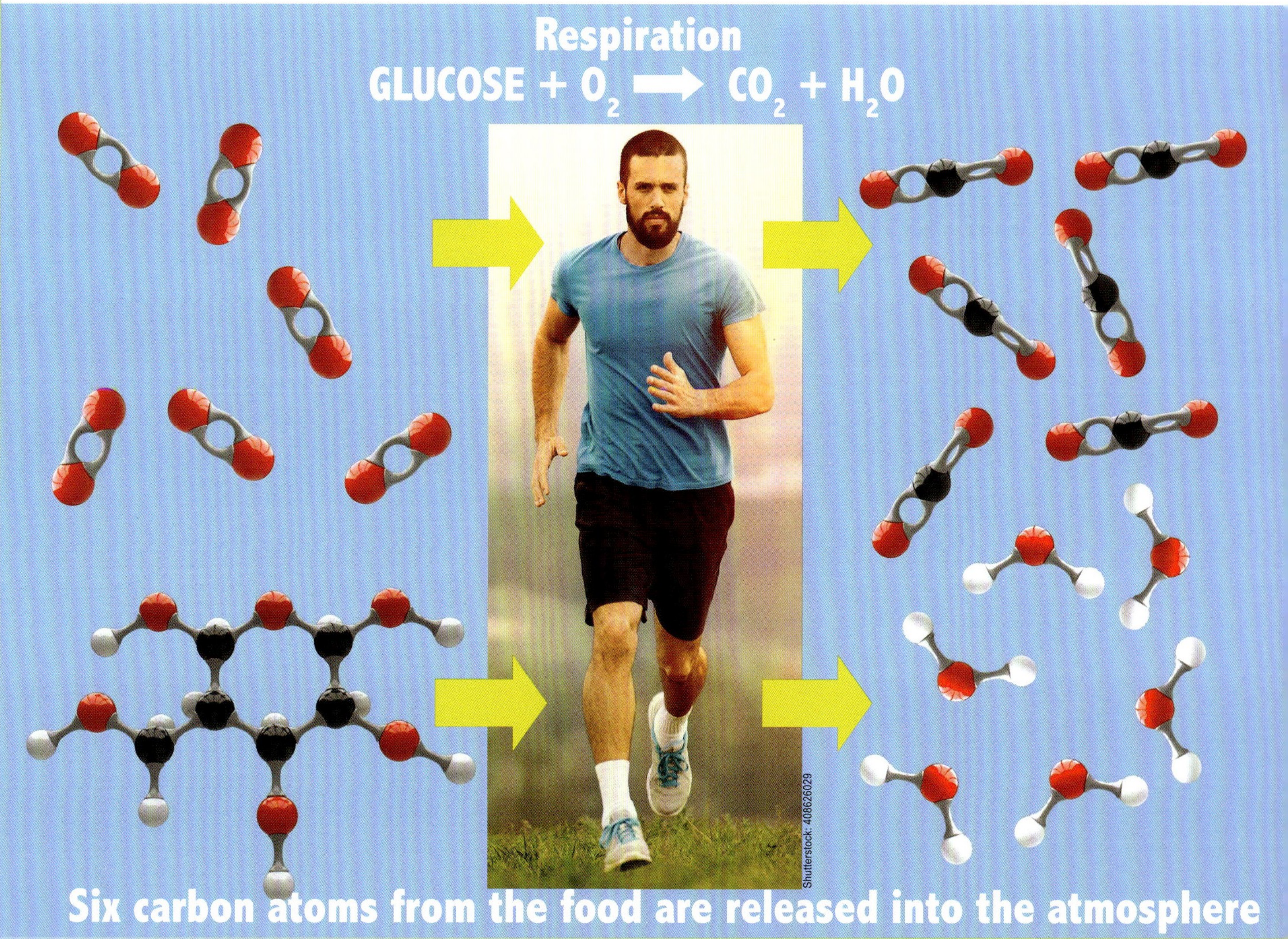

Respiration
$$\text{GLUCOSE} + O_2 \longrightarrow CO_2 + H_2O$$

Six carbon atoms from the food are released into the atmosphere

Carbon

Glucose is made by photosynthesis and used up in respiration. Every glucose molecule contains six atoms of carbon. When one molecule of glucose is used in respiration, six molecules of carbon dioxide are formed. The number of atoms taken in must balance the number of atoms given out.

Chemical reactions do not change atoms from one form to another. Nor do atoms get created or disappear. The total number of atoms on Earth does not change because of chemical reactions. In fact, the number of atoms on Earth has remained much the same since it first formed. At times some get added from space when a meteorite lands. As we saw a few pages back, that is usually not good news – just ask the dinosaurs.

Some atoms also leave Bubble Earth when we send rockets to the moon or planets, but the loss of atoms is a very, very tiny fraction of those left behind.

So the number of carbon atoms inside Bubble Earth is fixed. But that does not mean that the amount of carbon in the atmosphere remains the same. That only happens if plants can process all the CO_2 produced by animals and their activities. The diagram alongside shows this situation.

Three hundred million years ago the balance was heavier on the plant side of the see-saw. Back then, the land was covered by plants with very few animals to eat the food they produced. Fortunately there was a lot of CO_2 in the air from

A scene from three hundred million years ago when only amphibians and invertebrates lived on land.

volcanic eruptions that could be used for photosynthesis. Much of the food stored in trunks and leaves piled up in swamps, lakes and seas, where it was turned into fossils. Coal and oil are the remains of those fossilised plants. They are called fossil fuels and are very important in the climate-change story, because burning those fuels returns the carbon to the atmosphere.

Light

Photosynthesis needs energy in the form of light. Without light, life could not exist. This is the only thing Earth needs from outside the bubble.

Light is energy and must be understood in a different way to matter, which is made up of atoms. We use photons to help us understand the behaviour of light. A photon is a tiny bit of light that carries energy. The important thing about photons is that they only exist when they are moving. If they stop moving, they disappear. This means you can't collect photons in a bottle so you can use their energy later.

We will picture photons as screw-shaped arrows. The number of turns to the screw indicates the type of photon. The arrow shows the direction it is moving. Here are the photons important in understanding the climate-change story:

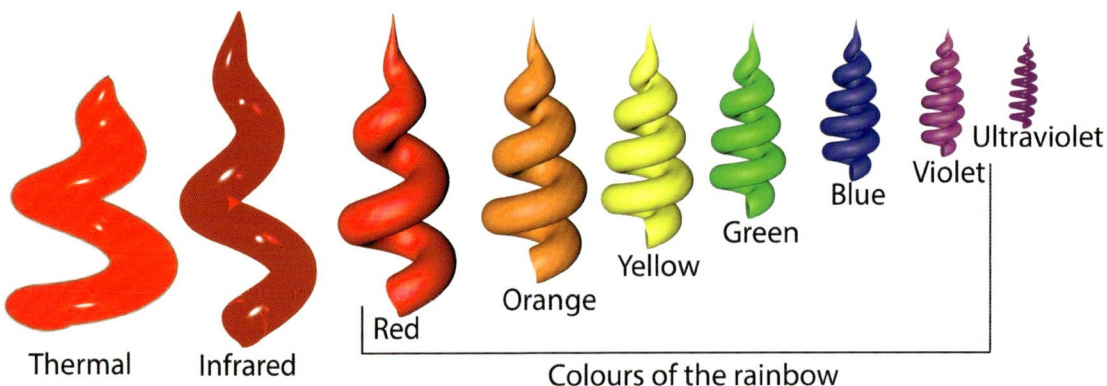

Our eyes can see colours-of-the-rainbow photons. And our skin can feel the heat of thermal and infrared photons. But we can't detect ultraviolet photons – the ones that damage our bodies unless we block them with sunglasses and hats.

Right: The boy's temperature is being taken by a non-contact thermometer which measures the energy of the thermal photons emitted by his body.

Below: The dog and human have been photographed with a camera sensitive to thermal and infrared photons. Notice that one of the coldest parts of the dog is its nose, which you've probably discovered if you've ever played with a dog.

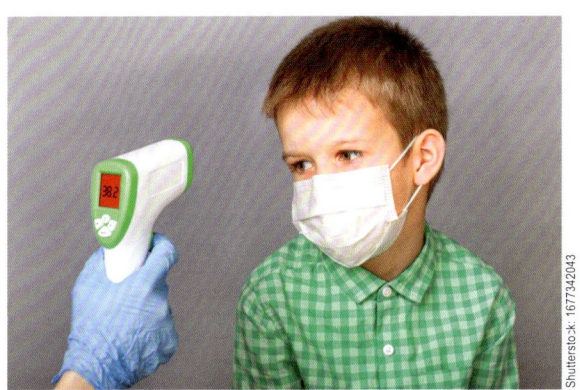

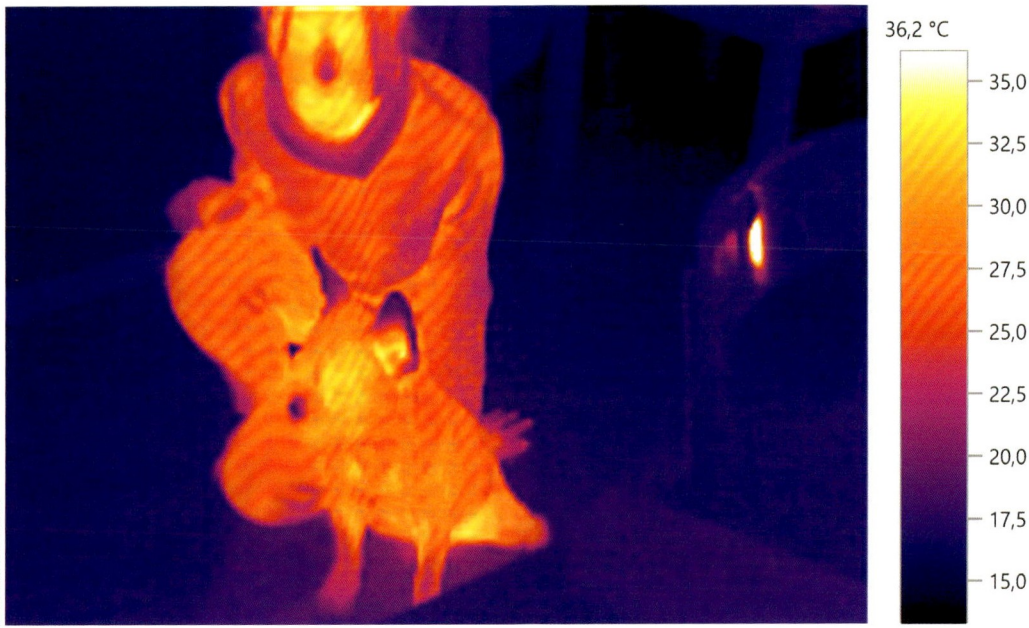

The sun generates a vast number of photons every second, which gives Earth most of its energy. But it is not the only thing to create photons. Everything does, and the hotter something is, the more are produced. Mostly we can't see these photons because they are thermal or infrared. But we have instruments that can detect them. An infrared thermometer can measure your temperature from the type of photons your body emits. Special cameras can convert infrared images into ones we can see.

Properties of Light

You will have seen cars with reflective shields blocking out the sunlight (top, left). You've probably also climbed into a car which hasn't had one and found how hot it was just from sitting in the sun. Studying this effect will help us understand why Earth's temperature is rising.

We'll start with reflection of ligh from a mirror. We think of sunlight as squadrons of photons ranging from infrared to ultraviolet. The image (centre, left) shows some of these squadrons arriving at the side mirror of a car. All the photons bounce off the mirror, except the ultraviolet (UV) which crash in the glass of the mirror. Anyone looking at the mirror from within the car would be blinded by so many photons entering the eyes and would need to turn away.

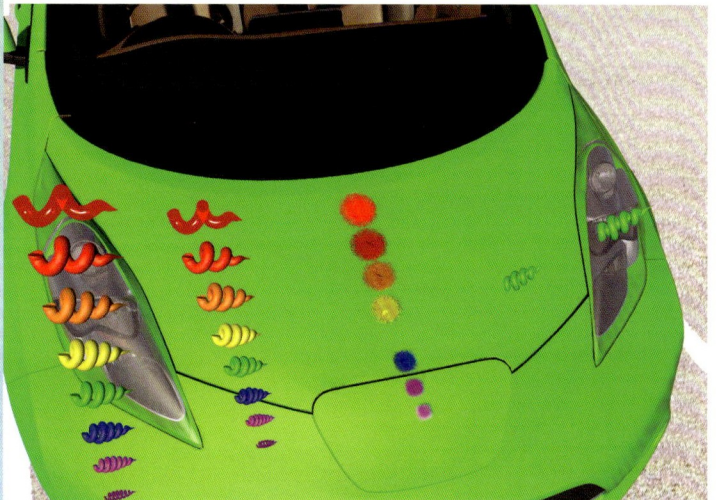

Now let's look at squadrons of photons hitting the bonnet of the green car (bottom, left). All except one type crash when they hit the surface. The ones that don't are the green photons. They reflect. That's why we see the car as green. If it had been a red car, then red photons would have been reflected when all the others crashed.

Crashed photons give up their energy to the surface. In this case, the bonnet will get hot, possibly so hot you can't hold your hand on it. If it were a green plant, some of the crashed photons – red, orange and yellow – would have given their energy for photosynthesis to make food and oxygen. The others keep the plant warm so it can grow faster.

If you hold your hand near the hot surface of a car you will feel the heat coming off. Your body is detecting thermal infrared photons. These are different to the infrared photons in sunlight and have different properties.

All objects give off thermal photons. The hotter the object, the more thermals are produced. This can be shown using a special camera that detects thermal photons rather than visible ones. The kangaroo above was photographed feeding at night when no visible photons were around. Thermal photons have been drawn to show how they leave a warm body.

Thermal photons can't pass through glass and some plastics. This can cause them to get trapped by windows, which is what happens in a car. Sunlight passes in through the windows with the photons crashing on the seats and fittings. The seats get hotter and emit more thermals. But the thermals can't pass out through the windows. They

bounce around, some crashing back into the seats, making them hotter still. The inside of the car becomes a cage for thermal photons and can reach temperatures over 40 °C. If reflective shields are used the light photons get reflected back out so that the car doesn't heat up quite so quickly. If you leave the windows open this allows the thermal photons to escape and slows the rate even more.

Greenhouse Effect

Thermals getting trapped in a place because they can't pass through glass or some other substance can be referred to as the greenhouse effect. A greenhouse is another name for a glasshouse or hothouse – the glass buildings used to grow plants in winter or where the climate is too cold for them to grow naturally.

Earth's atmosphere acts like the glass in a greenhouse, as shown in the image to the right. The scene is the Tasman Sea with New Zealand nearest, Australia to the left, New Caledonia and other Pacific Islands to the right and middle. Squadrons of photons are arriving from the sun that has risen a few hours before. Some types of photons, like green and blue, reflect back into space. But as you can see most of them crash into the land and sea, giving up their energy to heat the planet.

Like all things, land and sea emit thermal photons. Some escape into space, others get absorbed or reflected. The number that escape depends on the amount of carbon dioxide (CO_2) in the atmosphere. The more CO_2 the more thermal photons get trapped, making Earth warmer.

This greenhouse effect is a good thing. Without it, life as we know it could not exist on this planet, as the temperature would be too cold. Problems begin when the gases in the atmosphere change too quickly.

Water vapour in clouds also traps thermals. Fewer thermals escape when the sky is cloudy. This can be very obvious in winter, when we have frosts on nights if the stars are visible. That's because there are no clouds to trap the thermals and they escape. Without the thermals, the temperature can drop low enough for water to freeze, forming ice on the ground.

The gases that cause the greenhouse effect are called greenhouse gases. The ones causing the most problems are carbon dioxide, nitrous oxide and methane. We'll look at the two carbon ones in this story, as their effects and problems are the easiest to understand.

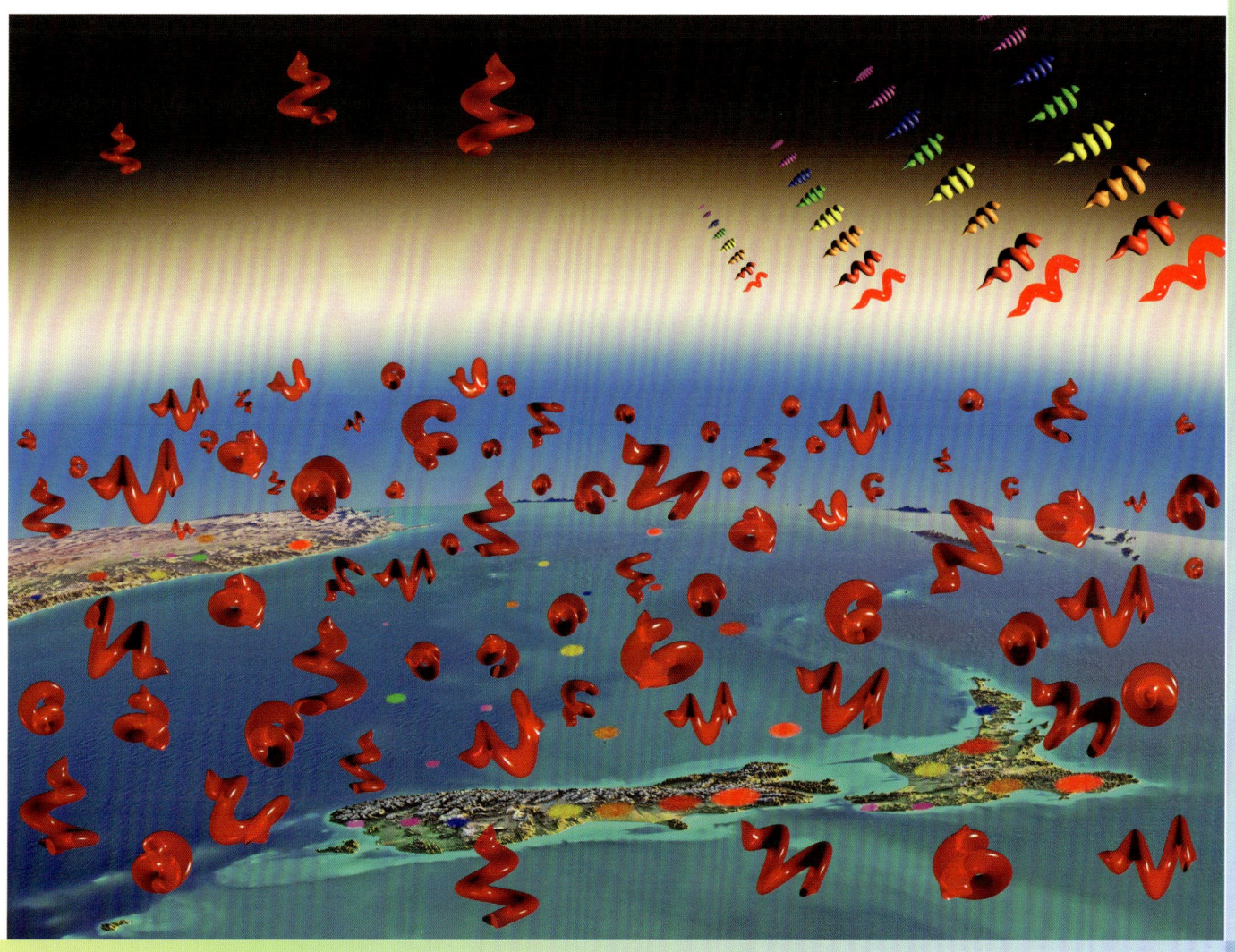

3 CAUSES & CURES

Fossil Fuels and Carbon Dioxide (CO_2)

Fossil fuels are the remains of living organisms that lived within Bubble Earth well before we humans did. Coal and peat are formed from plants that grew on land; petroleum oil and natural gas come from plankton and algae that floated around in the sea. Whatever the type of fuel, they are all a by-product of the photosynthesis reaction:

$$CO_2 + H_2O \Rightarrow \text{glucose} + O_2$$

Glucose is easily changed into the many other carbon compounds that make up plants and animals. When living things die, the carbon compounds change again. Under the right conditions they are turned into the molecules of fossil fuels. Every carbon atom in a fossil fuel was originally part of a carbon dioxide molecule in the atmosphere. So the formation of fossil fuels is really:

$$CO_2 + H_2O \Rightarrow \text{fossil fuel} + O_2$$

This process has been going on for around 300 million years and is important because it formed a lot of the oxygen that is now in the atmosphere. So for 300 million years, Earth has been storing carbon in fossil fuels using energy from the sun. It's gone on summer, autumn, winter and spring for as long as there has been life in Bubble Earth. Only in the last 200 years have we been digging up those fossil fuels and reversing the reaction using the combustion or burning reaction:

$$\text{fossil fuel} + O_2 \Rightarrow CO_2 + H_2O$$

Every time we use fossil fuels in our cars, trucks, factories, power stations, barbecues, heaters ... we return carbon atoms to the atmosphere, increasing the levels of CO_2.

We've been digging up coal and pumping oil for so long that none of us really think about the consequences. So let's do that now. Look at the pictures on the opposite page.

The first shows Bubble Earth orbiting the sun sometime way back in the past. The plants are going about their business of photosynthesis, passing through the seasons as they have done for hundreds of millions of years. In the process, they use energy from the sun to store some of the carbon atoms in fossil fuels. The illustration shows nine of those years.

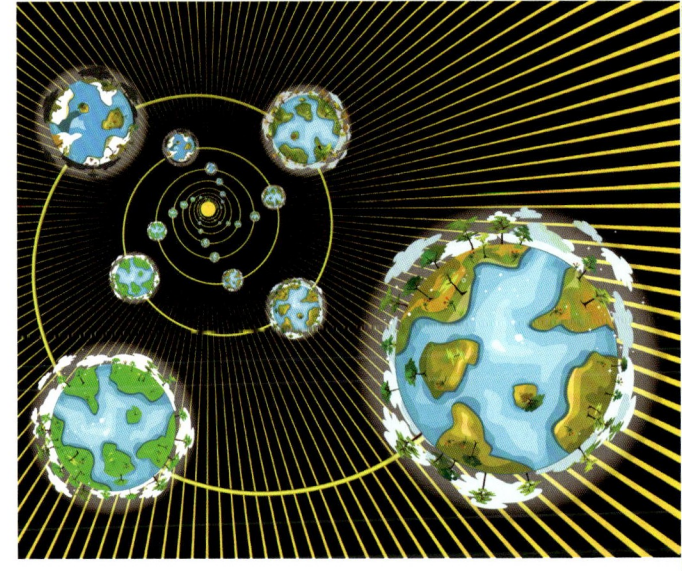

The second picture has the fossil fuels being extracted from the earth, which we have done on a large scale for the last couple of centuries.

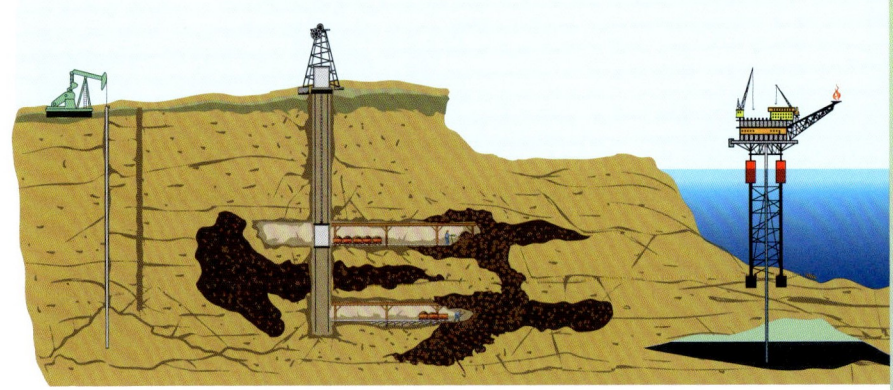

The third shows the fossil fuels being burnt in vehicles, factories, power stations, schools, offices … It's happening almost everywhere there are lots of humans. This process uses up oxygen and returns carbon dioxide to the atmosphere.

So how much fossil fuel can be formed during nine years of photosynthesis? That's longer than some of you reading this have been alive. As you can expect, it's a big number, one that few people can understand.

So, let's ask the question another way. If we were to use the fossil fuel from nine years of photosynthesis, how long would it last? The answer is two. But is that two years? Two months? Two weeks? Two days? Two hours?

None of these – it is just two minutes.

In 2020, the humans in Bubble Earth used up nine years of carbon storage in **two minutes**. That's probably about the time it takes to read this page. It means that to keep us in fossil fuels for the whole year required about two-and-a-half million years of photosynthesis!

Putting it yet another way: in one year we returned to the atmosphere all the carbon dioxide that took two-and-a-half million years of photosynthesis to remove. We can't keep behaving like this.

Returning the Balance

This see-saw shows the current state of carbon dioxide in the atmosphere during the first decades of the 21st century: there is much more carbon dioxide being put into the atmosphere by humans than can be taken out by plants. One of the problems is that there are too many humans, but there's not much any one of us can do about that. So if there is to be a change, it has to be in how much carbon dioxide each of us puts into the atmosphere. To get some idea of how we create CO_2 we'll look at an average day for a child.

You wake up and continue breathing in oxygen (O_2) and breathing out carbon dioxide (CO_2) as you have been doing all night. You're an animal – that's what you have to do to stay alive.

The first thing you do is get dressed. Every item you put on will have needed energy to make it and transport it to you. More than likely, that energy came from fossil fuels.

For most of us, breakfast will be something that had to be manufactured, packaged and transported. Which is still more fossil fuels going up in smoke.

Off to school. Whether you get transported by caregiver or school bus, this will be the most obvious use of fossil fuels in your day.

No matter what happens at school during the day, energy will be used to heat (or cool) and light your classroom.

Before bed, you might have a shower or bath using even more energy. Then it's back to bed. Remember to keep on breathing out CO_2, because that's what us animals must do to keep the plants alive.

Carbon Footprints

By now it should be obvious that holding your breath until your face turns blue isn't going to save you or Bubble Earth. But there are other things we can do, both as countries and as individuals.

In 2016, most of the countries in the world agreed to aim at keeping global temperatures no higher than two degrees above what they were in 1900. Already the temperature is one degree higher than 1900, so there's a lot of work to be done to stop it climbing up that second degree.

One way countries calculate how much carbon is going back into the atmosphere is by working out a 'carbon footprint'.

This is the amount of carbon used by our actions in one year – the sort of things shown in the boxes on the right of the see-saw. It is measured in tonnes of carbon dioxide, although we mostly forget about the unit and just use the number. Remember, this is a competition you don't want to win: bigger is not better.

The organisation that works with countries on climate change is a branch of the United Nations called the UNFCCC.

A carbon footprint can be calculated for a person, a family, a business or a country. In a developing country, such as many in Africa, the number can be less than one for each person.

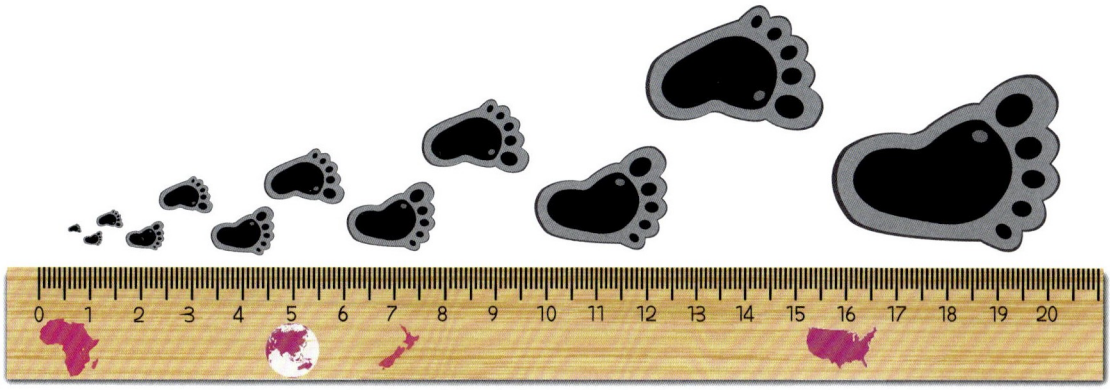

In a country such as America, which is using a lot of fossil-fuel energy, it can be as high as 16. New Zealand is between seven and eight. The world average per person is about five.

Measuring a carbon footprint is rather tricky. There are lots of apps and online calculators, but most are intended for adults and businesses. However, if you want to get some idea of the sort of questions they ask, you might like to try some of the links given in the Resources section on page 55.

Calculators also suggest ways of reducing our carbon footprint. While you might not be able to try all the ideas suggested, the important thing is to see if you can get your footprint as small as possible.

Agriculture and Methane

Most farm animals are herbivores, which means they eat plants. Plants are made by photosynthesis. Thus, all the carbon atoms in the body of a herbivore come from CO_2 in the atmosphere which make a herbivore carbon neutral when counting atoms: carbon atoms into the air = carbon atoms taken out. This *should* mean that herbivores don't contribute to global warming, but it doesn't. The problem is that the carbon they put into the air isn't all CO_2. A lot of it is methane – CH_4 – and methane is a greenhouse gas that is a hundred times more effective at trapping the thermals than carbon dioxide. So when working out the carbon footprint, methane has a count of 100 compared to carbon dioxide's one. Methane causes a hundred times the warming of CO_2.

Methane is produced during digestion of plants in all large animals but is particularly bad in ruminants which includes cows and sheep. It is passed out of the body via eructation and flatulence – otherwise known as burping and farting. The diagram (opposite page) shows the carbon cycling in a cow.

If the amount of methane could be reduced, New Zealand's carbon footprint would become much smaller. Food supplements give the greatest hope and one of these is a natural product. The red algae shown below is a seaweed which is commonly called harpoon weed – *Asparagopsis armata* to a scientist. It is a native to New Zealand and Australia which, over the last century, has been accidentally spread all around the world by shipping. In some countries it is grown in sea farms for use in cosmetics and food supplements. One of its properties is slowing the growth of bacteria. If fed to cattle and sheep, it reduces the amount of methane produced by bacteria in the gut.

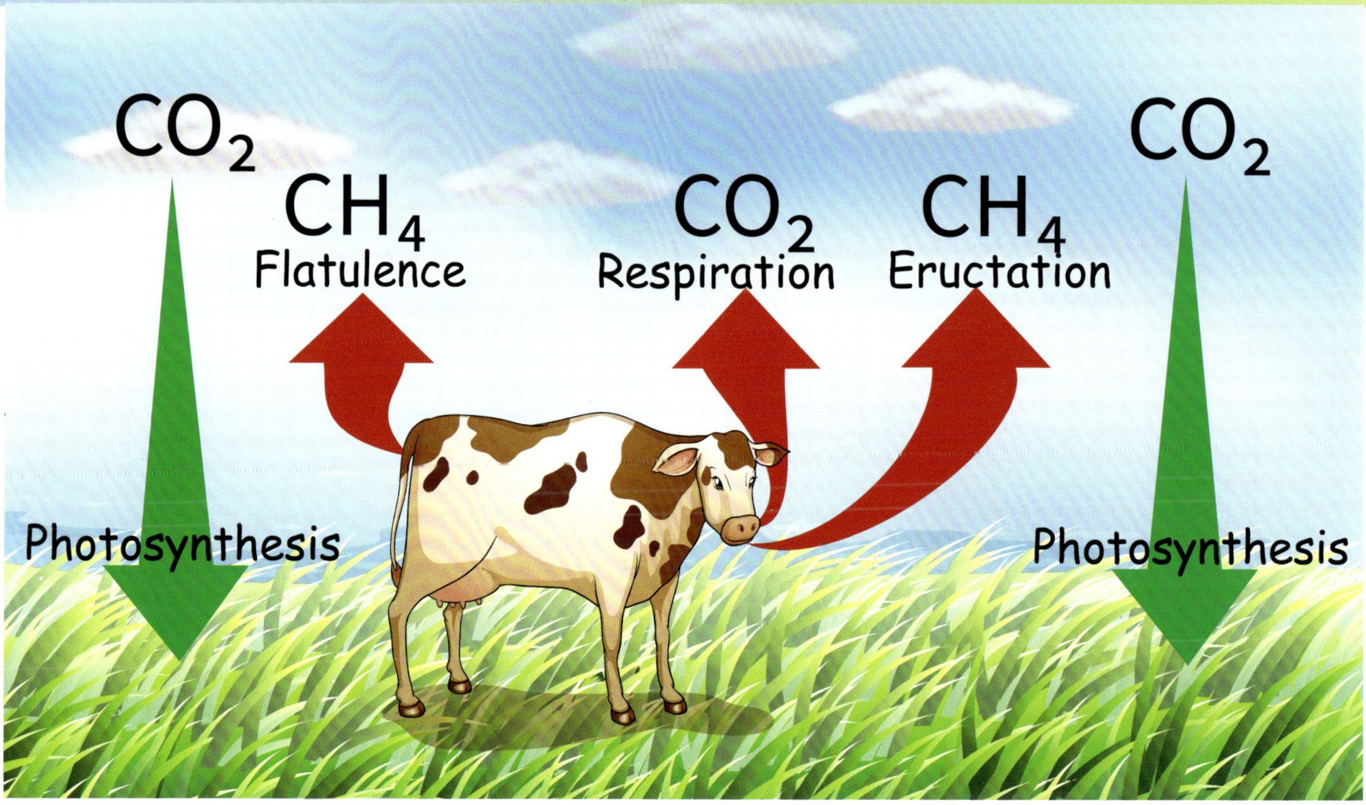

All mammals that eat plants are likely to produce methane, including humans. This becomes a problem when treating our sewage, where large amounts of methane are produced. This has to be carefully managed so it doesn't catch fire. The best way is to collect the methane and use it as a fuel just like natural gas, which is mostly methane. This way the methane can be used to produce electricity. Sure, the carbon gets released as CO_2, but that is a hundred times better than the CH_4 getting out into the air.

Harpooon weed – Damsea.

Methane producing and powered sewage plant

Zero Carbon

Hugging a tree has often been an image used for people who care for the environment – and calling someone a tree-hugger is sometimes used as an insult. And yet we should all be embracing trees because they help us keep the temperature down.

All plants do photosynthesis, but trees are the best at taking CO_2 from the atmosphere and locking the carbon away for a long time. Smaller plants, ones that live for just one season, quickly return the CO_2 when they rot at the end of that season. With a tree, the carbon can be trapped in the wood for thousands of years.

The figure alongside is carved out of New Zealand swamp kauri. The tree was probably a thousand years old when, somehow, it fell into swampy ground. There it lay for up to 60,000 years before being dug up. Even now that it's exposed to the air, the wood is so hard the carbon could be trapped for thousands more years.

Forests are important in our attempts to keep the global temperature down. The idea is to add more blocks to the left side of the see-saw shown on page 30. This is the thinking behind schemes such as Zero Carbon and Carbon Neutral, where organisations balance the use of fossil fuels by planting trees. Let's see how they work.

Deforestation in Brazil, South America

We'll assume your carbon footprint is around seven. To remove all that carbon from the atmosphere requires 300 pine trees. These would take up the space of about half a football field. For your whole family it could be several football fields. For all the carbon used by a city, the forests would need to be at least 20 times the size of the city, which raises the question: Is there enough spare land for all these forests?

Possibly – at least for 30 years or so. Hopefully, that's long enough for us to find better ways to stop carbon getting into the atmosphere in the first place. Anyway, planting more trees is much better than what is happening at present, where forests are being cleared to create more farmland to feed the world's increasing population. This deforestation is just adding to the problem.

Renewable Energy

Sunlight is the original source of almost all the energy we use. That's not surprising, because sunlight is the only thing that comes into Bubble Earth from outside. Fossil fuels were formed by sunlight; wind turbines are turned by winds created by sunlight; hydroelectric power stations depend on rain which comes from clouds that formed because of sunlight. The only energies that can't be linked to the sun are nuclear and geothermal. They come from the radioctive uranium inside Earth, that has been there since Bubble Earth first formed.

A renewable energy is one that can be used without fear of it ever running out. Let's look at the six main types.

Solar
Panels absorb the energy of crashed photons and convert it directly to electricity. They can be placed on the roofs of buildings, or in large arrays called solar farms. A big one in Auckland floats on a sewage-treatment lake.

Hydroelectricity
Dams collect river water, taking it through turbines to generate electricity. New Zealand gets more than half its electricity this way. The main problem with this system is that the dams often flood a lot of land, which interfere with the environment.

Wind
New Zealand is a hilly country on a windy part of Earth's globe. Wind power produced by turbines could eventually provide almost half of our energy. There are some problems with the noise the turbines make, and they do alter the look of our natural landscape.

Geothermal
The inside of Earth is hot, particularly around volcanic areas. The best way to use this energy is to pump water down and use the heat it brings back up. This is used to heat homes in Rotorua and make electricity around Taupō.

Waves and Tides
Waves and currents in the ocean are formed by the sun's heating combining with the gravitational pull of the Moon and Sun. Ocean-based power stations are difficult to build and can interfere with the environment.

Biomass Fuels
Biomass is anything that is, or was once, living. It can be made into fuels such as ethanol, methane and hydrogen. The advantage is that the fuel can replace fossil fuels. The problem is that, unlike all the other renewables, it puts CO_2 into the air.

Hydrogen and Fuel Cells

Electricity is the key to reducing carbon dioxide in the atmosphere. As you can see from the previous page there are lots of ways we can make electricity without releasing greenhouse gases. The big challenge is how to use electricity in transporting people and things. This means changing all the trucks, ships, planes and cars that now use fossil fuels over to electricity. Let's see how this might work.

There are three ways an electrical vehicle (EV) can get its electricity: make it using solar panels on the roof, like the cartoon cart above; connect to power lines, as electric trains do; use batteries.

The battery method is the only one that works night and day, and on any road. The main problem with batteries is the time it takes to recharge them. An overnight charge will cover most of a day's travel to work and around town, but not a longer trip. A petrol-powered vehicle needs just a few minutes to top up the tank; with batteries it takes a few hours. That's a big inconvenience.

There is, however, a type of battery that overcomes this problem. It's called a fuel cell. The best of these use hydrogen as a fuel. Hydrogen can be made by passing electricity through water.

$$H_2O \Rightarrow H_2 + O_2$$

The reverse reaction happens in a fuel cell, giving the electricity back, as shown in the diagram on the following page.

So an EV will have a tank of hydrogen and use oxygen from the air to produce the electricity for the motor. When the tank is getting empty, the driver will pull into a gas station and refuel. The great thing about this system is that the only substance put into the atmosphere is water (H_2O).

Hydrogen fuel cells can also be used in planes as they are much lighter than rechargeable lithium batteries. In 2020, the largest electric plane could carry 10 or so passengers. By the middle of the century, it is expected they'll be big enough to replace many fossil-fuelled planes.

In the meantime, there is another way hydrogen could help airlines lower their carbon footprint. Hydrogen can be reacted with biomass to form kerosene, the fuel of most airliners. Yes, the carbon of the biomass will be poured back into the atmosphere, but that's still much better than burning a fossil fuel, and it's technology we can use *right now*.

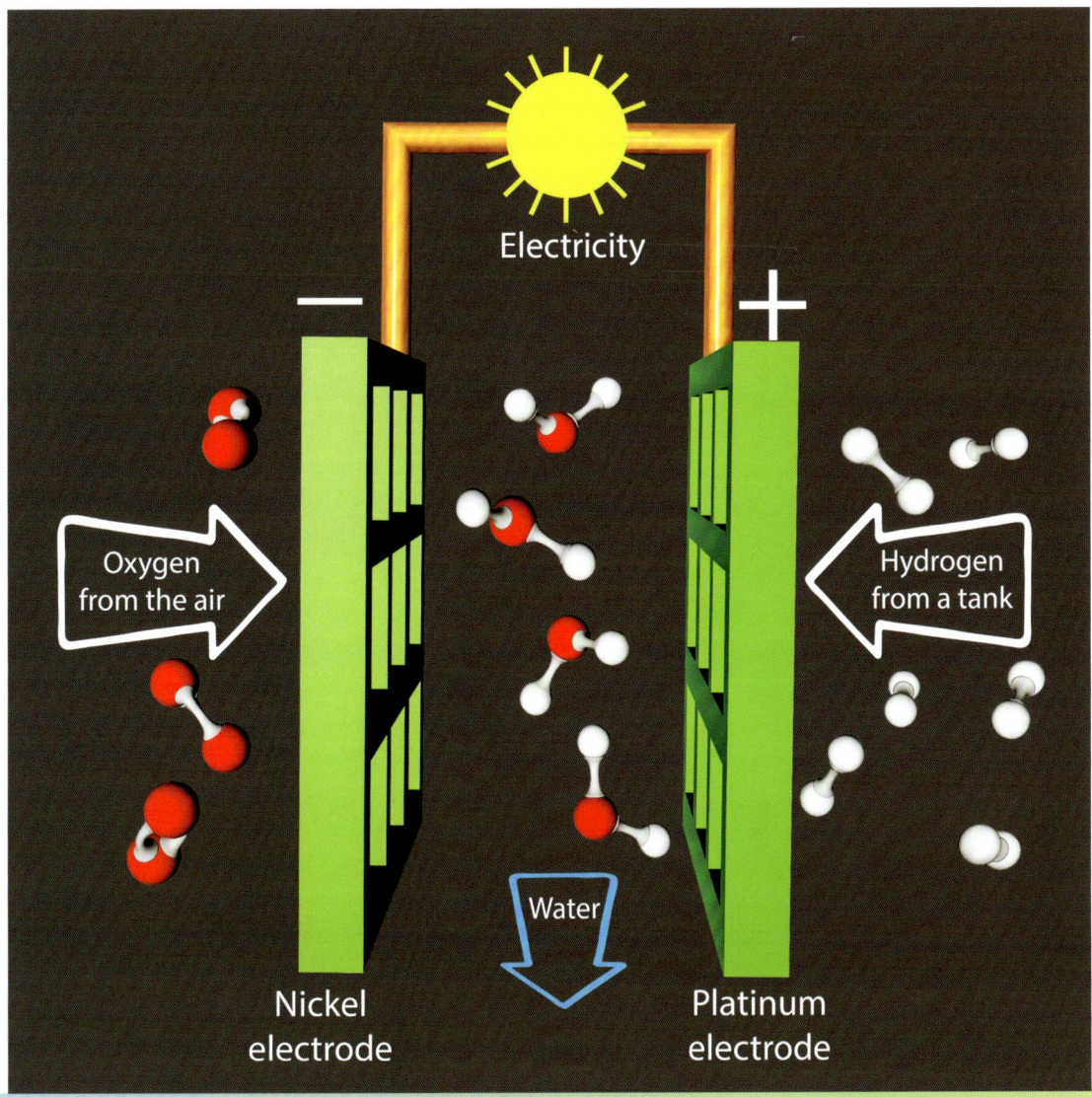

These two views of Franz Josef glacier on the South Island's West Coast were taken nine years apart, the one on the left in 2006 and the other in 2015. In that time the leading edge of the ice has retreated more than a kilometre up the valley.

4 CONSEQUENCES

Now is the time to talk about the scary stuff. This is where we'll look at what is likely to happen over the coming years even if we do all the things that should reduce greenhouse gas emissions.

The first thing to know is that change of climate doesn't happen instantly. We see that in the normal yearly seasons. In the southern hemisphere, the sun is highest in the sky and the days longest just before Christmas, and yet the hottest weather comes a month later. The same with winter: the shortest day with the sun lowest in the sky is 21 or 22 June, but most of the snow and icy weather comes in the following months.

The main reason for this is that the oceans take a long time to heat up, and an equally long time to cool down. We can expect a similar lag with global climate. If we do manage to stop the temperature rising, we probably won't see the results straight away. It may take years for any change to become obvious, and things could get even worse in the meantime.

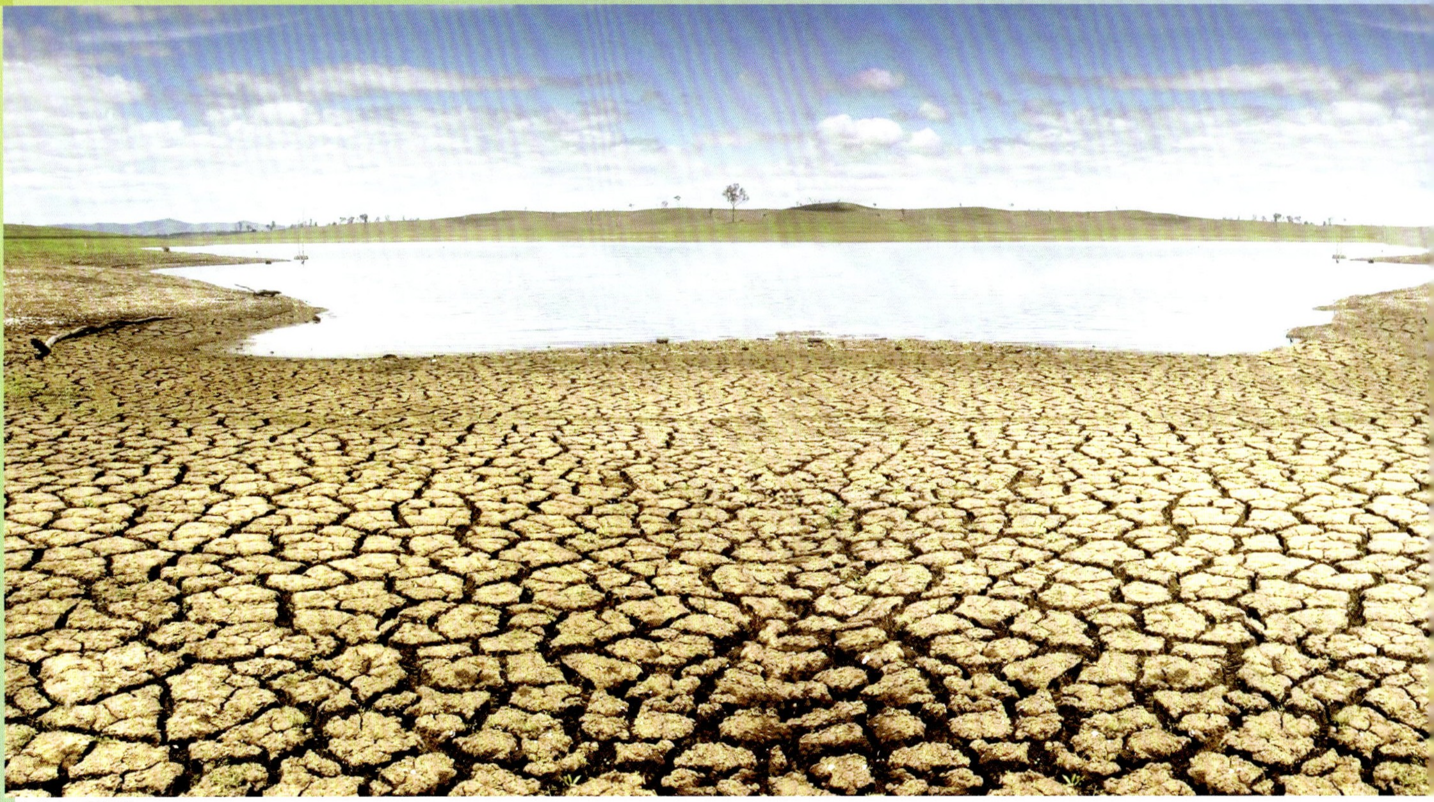

Shutterstock: 96114200

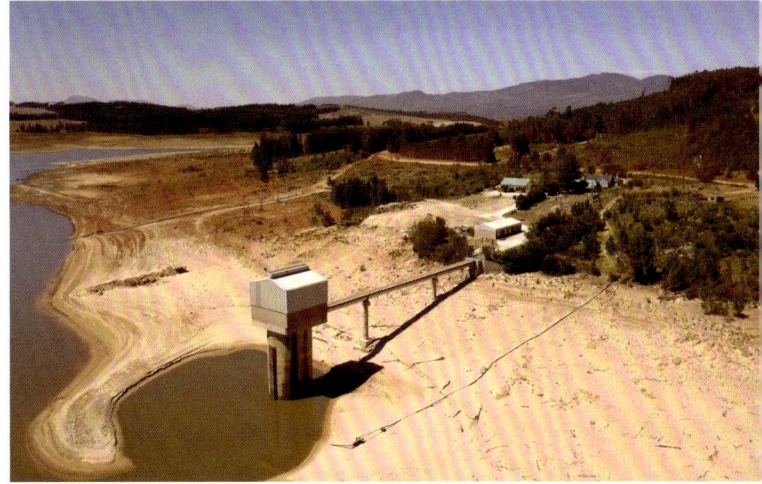

Extreme Weather Events

Rising global temperatures make the water in the seas warmer. Warm water evaporates more quickly than cool water. More evaporation means more clouds; more cloud means more rain for some places and a greater chance of flooding. Already many parts of the world have seen downpours worse than any previously recorded. New Zealand experiences many floods a year, and there is some evidence that they are happening more often with the water levels higher than expected.

Weather patterns are hard to predict, but one thing is becoming clear: as some places are getting wetter, others are getting drier. In the dry places, water supplies are lower than ever before, farm pastures are browner, and bush fires happen more often. Water is becoming a precious resource which we will need to manage in different ways if we want to continue to live in the same places. No living thing, animal or plant, can survive without water.

Water is also causing big problems on the coastlines. Like all liquids, water expands as the temperature increases. This means that as the seas get warmer, they take up more space – the sea level rises. So far the sea level has increased 20 centimetres since 1900. A good bit of this increase is due to ice caps melting in mountains and polar regions. While 20 centimetres doesn't sound much – it's less than the height of this page – when combined with storm waves and high tides, the sea can advance a long way inland, destroying crops and buildings. This is a particular problem for coral atoll islands in the Pacific and Indian oceans. Countries like Kiribati and Tuvalu, both with strong ties to New Zealand, are already anticipating there will be a day when they must leave their islands and find another place to live.

© Stuff NZ

Melting of ice caps doesn't just raise sea levels, it also raises the temperature. White ice reflects almost all the sunlight reaching it, sending the photons back into space. They act like a shield put in the windscreen of a car (see page 24) to keep it cooler. The smaller the ice caps, the hotter Bubble Earth becomes.

© Stuff NZ

Ocean Acidification

Bush fires, floods and storms are events that happen quickly, with the damage becoming obvious soon afterwards. This is not so with other disastrous consequences of climate change. Sometimes, the effects may not be obvious until it is too late to do anything about it. Here we will look at one called acidification.

When carbon dioxide dissolves in water, it forms an acid called carbonic acid. The reaction is:

$$H_2O + CO_2 \Rightarrow H_2CO_3$$

This reaction helps many sea creatures because the CO_3 part can be combined with calcium atoms – Ca – to make hard shells. The substance is called calcium carbonate, with the chemical formula $CaCO_3$. Calcium carbonate is a common compound found in many rocks. One of these, called limestone, is formed from the ancient shells of sea animals. In some limestone rocks, you can actually see the shells as shown in the photo below. In most, however, the shells have been pressed and heated so much that it is difficult to tell that the substance was once the shell of a living animal.

Calcium carbonate dissolves in carbonic acid. This is how limestone caves are formed, such as those found in many places in New Zealand. But it also

means the shells of sea animals can dissolve while they're still living. The more carbon dioxide in the atmosphere, the more carbonic acid in seas and lakes and rivers. Therefore the more likely it is that shells will dissolve. The snail on the right, is one that floats on the ocean surface. You can see that it has a thin see-through shell; this will only get thinner as CO_2 in the atmosphere increases.

Just about all shelled sea animals spend the start of their lives floating in the sea as plankton. This is when the shell starts developing and is soft with little resistance to acid. If they can't form a solid shell they will die, and so might the animals that depend on them for food, including humans.

Coral is also calcium carbonate, which means coral reefs are attacked by acidification. New reefs will have difficulty forming and older ones will be weakened. Take out the corals, and many of the other animals will disappear as well. Many species are likely to become extinct.

Humans will be affected, too. The death of a coral reef severely influences those of us who live on coral islands. Reefs, and the lagoons they form, are a major source of food. They also provide a barrier that protects the islands from the worst ravages of storms – yet another consequence of climate change affecting the small island nations of the Pacific.

Living coral

Coral killed by acidification

Mass Extinctions

A mass extinction involves the death of about three-quarters of all species in existence across the entire Earth over a short geological time. There have been five of these events during Bubble Earth's long history. While the triggering event has varied from volcanic eruption through to asteroid impact, the resulting changes in the atmosphere are what turned it into a mass extinction event.

The cause of a possible sixth mass extinction, the one happening now, is a combination of several things, but they can all be traced back to what we humans are doing. This extinction is called the anthropocene extinction. The anthropo- start to the word means 'to do with humans'. For example, the adjective anthropogenic means caused by human activity.

What we have been looking at in the last few pages is anthropogenic climate change. There are other factors involved, such as slow changes in Earth's orbit around the sun, and minor shifts in the tilt of Earth's axis. But these changes are very slow and do not explain the current rapid change in temperature.

We cannot put the blame on some other event. The anthropocene extinction is because of what we humans do. It will be the only mass extinction caused by one of the species.

Being the smartest of all species, you'd think we would have realised what we were doing earlier. Well, we did. A Swedish scientist, Svante Arrhenius, wrote about anthropogenic climate change at the start of last century. But it wasn't until 70 years later that non-scientists began to discuss the problem. And we're still discussing it 50 years on. Unfortunately, that's about all some of us are doing – talking and arguing about it.

The decades from 2020 to 2040 are when we need to act before the temperature gets to the no-going-back level. For centuries we've been treating Bubble Earth as if we're the only ones living in the bubble, to the point where most of the other species would be far better off if we weren't here. It is time to realise our responsibilities and to start behaving as the most intelligent species there has ever been. Now is the time to start saving Bubble Earth before it becomes much too late. So far, the extinctions haven't reached the level of a mass event, but unless we change our ways, they most certainly will, and maybe within *your* lifetime.

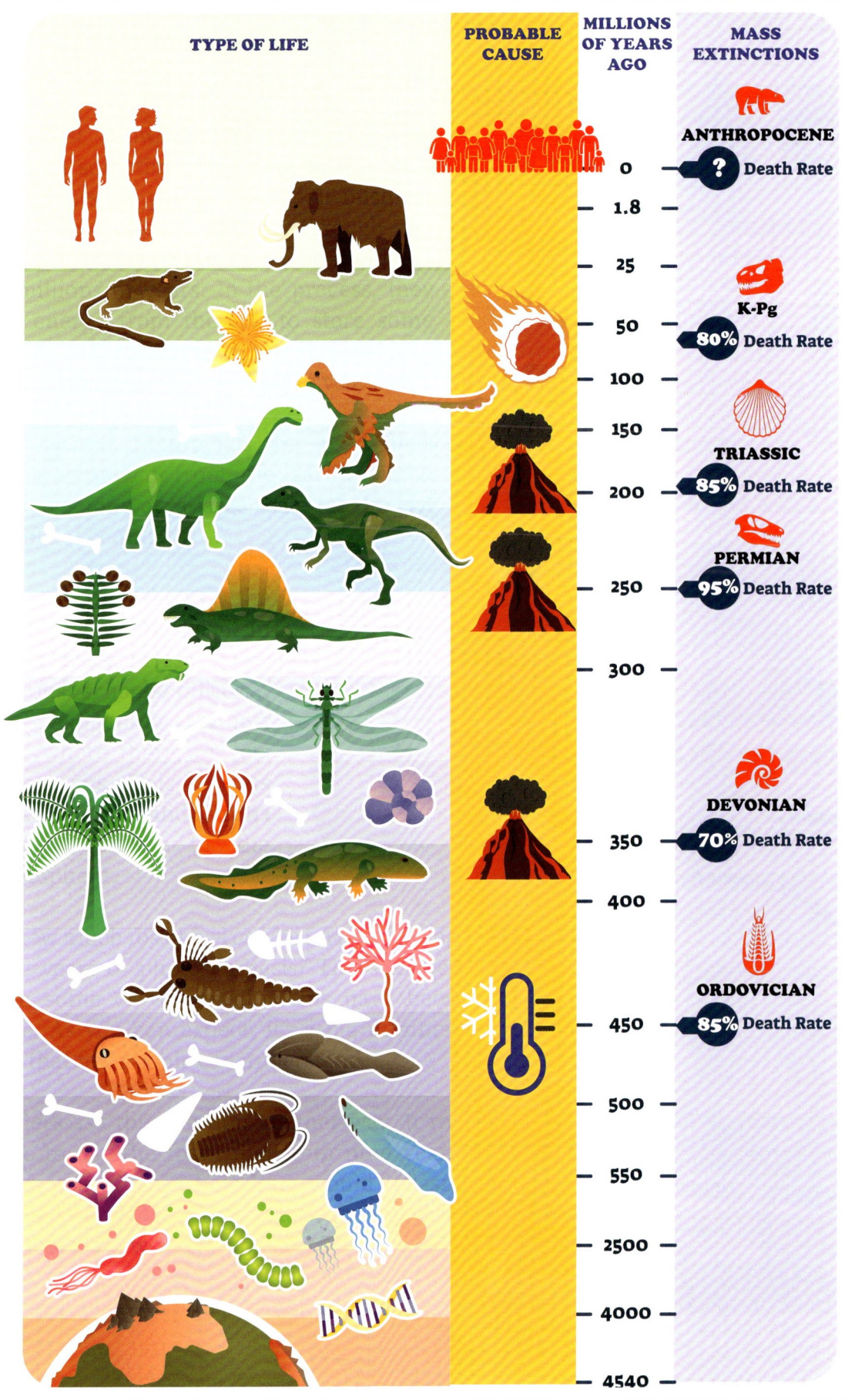

5 AT THE END...

Personal Responsibility

Finger-pointing and yelling at each other rarely achieves anything. What's needed is action. Many of us can't take action in a big, global way. The best most adults can do is to vote for governments that can and will do things. But children don't even get that chance. The thing all of us can do, however, is make sure we understand what is happening, and do the little things right – big movements often have tiny beginnings.

A good start is to use the four Rs: Reduce, Reuse, Recycle, Rethink.

Reduce Energy Use

Shut doors, have shorter showers, switch off lights and electrical devices, air-dry dishes, wash clothes using cold water. If it's winter, then close curtains early in the evening. Make sure taps aren't left dripping, especially hot water taps.

Reduce Waste

Buy products with as little packaging as possible. Look for those that use containers made from trees, instead of plastics made from fossil fuels. Be careful how you get rid of packaging because rubbish trucks and street-sweeping machines use fossil fuels. Do not burn rubbish; composting is far better. While both return carbon to the atmosphere, composting does it more slowly and can be used as plant fertiliser.

Reduce Consumption

Remember that everything you wear or use was manufactured using energy. If it came from another country, fossil fuels would have been used to get it to you. Do you really need to have the latest fashion clothes? If the device you've had for years still works, do you need to change it for the latest model?

Reuse

When you do need to get something new, think about how your old item might be reused by somebody else. Think: op shop, clothing bin, garage sale, auction site, roadside stall … what you consider junk may be someone else's treasure.

Recycle

The easiest materials to recycle are paper, glass, aluminium and steel. Plastics are much more difficult and many of them aren't recycled in New Zealand. Check the recycling symbols on stuff you buy. Those with numbers one and two (shown below) are recycled in New Zealand, but most of the others have to be shipped overseas using still more fossil fuels – and many are just dumped into landfills.

Rethink

This is the most important R of them all. Somehow we humans have to rewire our brains so that our impact on Bubble Earth becomes part of our everyday thoughts, not just something we think of every once in a while. Fortunately, there is still time to prevent the destruction of our planet, as long as we start doing it right now.

Activity: Grow Plants

By now you should understand that plants can help us get out of this climate-change mess.

A good way to learn about plants is to grow some. At the very least you will be taking carbon dioxide out of the atmosphere for a while. It might only be a small amount, but it is a start. We'll grow peas because: (1) they grow in a wide range of climates and (2) they're tasty even when uncooked.

Most plants begin life in a bubble called a seed. In these bubbles the tiny plants can survive for years before being released into a larger environment. They give us a chance to look inside a living bubble.

The most obvious part of a pea seed is the tough outer covering that protects the important bits inside. The scar visible in the photo is where the seed was attached to the parent plant. It is like your belly button: all that's left of a connection that supplied everything needed by a developing living organism.

Above the scar is a tiny hole. This is how the plant can breathe while in the seed bubble – oxygen comes in, and carbon dioxide goes out.

Opening the hard covering is not easy unless the seed is soaked in water for at least a day. When you do get inside, the seed will separate into two main parts, just like roasted peanuts or cashew nuts. These halves are the food supply for the plant while it develops.

Between them is the embryo, a term used for all living things before they leave their development bubble.

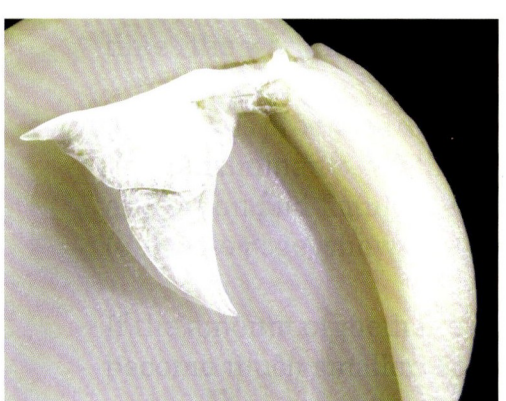

The bottom photo shows the embryo in detail: the two tiny leaves will develop into the green plant where photosynthesis occurs; the thick rounded part becomes the root which absorbs the water needed for photosynthesis.

For the pea embryo to develop further it needs water and oxygen. The photograph alongside shows a set-up we can use to study this change. To do this you'll need a jar, a paper towel, soil, seeds and of course water.

You can probably fit four seeds in a jar, but the packet will contain many more than that.

The rest can be planted in a pot or garden using the instructions on the packet. Put all the seeds in water and let them soak overnight.

Coil the paper towel into a cylinder so that it can fit into the jar. Spread it around so that it touches the glass. Trim with scissors if needed. Half fill the paper cylinder with soil. Put in the seeds on the glass side of the paper. Top up the cylinder with soil. Don't pack the soil too tightly as it will make it harder for the seed to grow.

Add enough water for the soil to be moist but not soaking wet. Rest the lid loosely on the top, taking it off during the days to let air circulate. Add more water as required. After a few days the plant will show obvious signs of life, as you can see in the photograph above. The ones in the garden may take 7–10 days before anything shows above ground.

Peas are climbers, so in the garden you will need to provide support such as wire netting or sticks. Your crop should be ready to eat within a few weeks. When the crop is finished, the vines make good compost or mulch for your next crop. Happy eating and gardening, and thanks for helping to save Bubble Earth.

GLOSSARY

acidification	the ocean becoming more acidic because of increases in atmospheric carbon dioxide
algae	plants without leaves and stems that mostly live in water; examples are seaweed and slime
amphibian	a cold-blooded animal that begins life in water breathing through gills which later change to lungs so it can live on land
animal	any living thing that obtains its food by consuming other living things
anthropogenic	caused or made by humans
asteroid	a small rocky object orbiting the Sun
atom	the building block of all matter
bacteria	tiny living things composed of a single cell
biomass	living material used as a fuel; examples are wood, energy crops like canola, industrial and human waste
bird	a warm-blooded animal that has a backbone, feathers, and lungs
calcium carbonate	chemical formula $CaCO_3$; the main substance in the shell of many animals such as seashells, crabs, snails and crayfish
carbohydrate	a food substance composed of atoms of carbon, hydrogen, and oxygen
carbon dioxide	chemical formula CO_2 – a colourless, odourless gas with molecules made up of two atoms of oxygen joined to an atom of carbon; makes up 0.04% of the atmosphere
carbon footprint	the amount of carbon dioxide released into the atmosphere by an individual, organisation, or community
carbon neutral	balancing every carbon dioxide molecule put into the atmosphere by taking one out, usually by planting trees; other greenhouse gases must be multiplied by a factor (100 for methane) to get the equivalent number of carbon dioxide molecules to use in the calculation
carbonic acid	acid that forms when carbon dioxide dissolves in water
cells	the building blocks that make up all living things
chlorophyll	the green substance in plants that absorbs light energy to power photosynthesis

climate	weather, such as rainfall, temperature and winds, averaged out over several years
combustion	a chemical reaction that uses fuel and oxygen producing water vapour and carbon dioxide; also called burning
COVID-19	a virus that infected humans in most parts of the world in 2020
deforestation	cutting down a forest and not replanting it
digestion	the processing of food in the gut of an animal
DNA	deoxyribonucleic acid, the molecules that contain most of the information that defines a species and the individuals within that species
embryo	the developing young of an organism before it has hatched, been born, or germinated
eructation	gases passed out of the gut of a mammal through the mouth; burping
fish	a cold-blooded animal with a backbone, scales, and gills
flatulence	gases passed out of the gut of a mammal through the anus; farting
fossil fuel	a fuel formed in the geological past from the remains of living things; coal, peat, oil, and natural gas
fuel cell	battery that converts the energy of a fuel into electricity
geothermal energy	renewable heat energy created by radioactivity within the Earth
glucose	chemical formula $C_6H_{12}O_6$ – the molecule that is the building block of all sugars and carbohydrates
greenhouse effect	energy is trapped within Earth's atmosphere by gases such as carbon dioxide, water vapour, and methane; this keeps the temperature on the surface warm enough for life to survive
greenhouse gases	gases in the atmosphere that absorb thermal energy; the main ones are carbon dioxide, water vapour, methane, nitrous oxide, and ozone
herbivore	an animal that gets its energy from eating plants
hydro energy	renewable energy of water flowing downhill
hydrogen	chemical formula H_2 – a colourless, odourless gas with molecules made up of two atoms of hydrogen; there is no hydrogen in the atmosphere

ice age	a time lasting tens of thousands of years when Earth's climate is much colder than usual; the last Ice Age ended 12,000 years ago
infrared	photons we can't see but can feel as heat; a special type is thermal infrared which is emitted by all objects; abbreviated IR
invertebrate	an animal that doesn't have bones such as a worm, shellfish or insect
mammal	a warm-blooded animal that has a backbone, lungs, and feeds its young on milk
mass extinction	a large number of different creatures becoming extinct over a short geological time
methane	chemical formula CH_4 – a colourless, odourless gas with molecules made up of four atoms of hydrogen joined to one atom of carbon; it is found in natural gas and formed during the decay or digestion of plants; a greenhouse gas that traps 100 times more thermal photons than carbon dioxide
molecule	a group of atoms that forms a particular substance such as water or carbon dioxide
nitrogen	chemical formula N_2 – a colourless, odourless gas with molecules made up of two atoms of nitrogen; makes up 78% of the atmosphere
nitrous oxide	chemical formula N_2O – a greenhouse gas that traps 300 times more thermal photons than carbon dioxide
oxygen	chemical formula O_2 – a colourless, odourless gas with molecules made up of two atoms of oxygen; makes up 21% of the atmosphere
ozone	chemical formula O_3 – a form of oxygen containing three atoms instead of the usual two; found in the top layers of the atmosphere
photon	can be thought of as the building block of light; photons have no mass and only exist when they are moving
photosynthesis	the process in plants where carbon dioxide is joined with water to form glucose; photosynthesis gets it energy from light
plankton	small living things drifting and floating in oceans and lakes
plant	an organism that makes its own food through the process of photosynthesis

Priestley, Joseph	a scientist who investigated the gases in air and the way plants and animals used those gases
protein	a food substance composed of atoms of carbon, hydrogen, oxygen, and nitrogen
renewable energy	a source of energy that cannot run out when we use that energy; examples are wind, solar, hydro, and geothermal.
reptile	a cold-blooded animal with a backbone, lungs, and scaly skin
respiration	a reaction inside a living cell where glucose joins with oxygen to give carbon dioxide and water; respiration is the reaction that provides the energy for all living things
ruminant	mammals that chew their food a second time after it has been partially digested; cattle, sheep, goats, antelopes, deer, giraffes, and their relatives
sewage	collected human waste such as faeces, urine, and washing water
solar energy	renewable energy from the sun
tidal energy	renewable energy created by the tidal pull of the Moon and Sun on the oceans of Earth
tuatara	Sphenodon punctatus, a reptile native to New Zealand and nowhere else; a survivor of an ancient group of animals that was common more that 200 million years ago
turbine	machine that is turned by a stream of water or steam
ultraviolet	photons with more energy than violet photons; we can't see or feel them but they can damage our skin; abbreviated UV
variation	difference between individuals of the same species; genetic variation is the differences in the DNA between individuals of the same species
vertebrate	an animal with bones such as a fish, reptile, amphibian, bird or mammal
water vapour	tiny droplets of water in the atmosphere that act like a gas
water	chemical formula H_2O – a colourless liquid with molecules made up of two atoms of hydrogen joined to an atom of oxygen
wind energy	renewable energy from the force of wind

FURTHER READING

General
A wonderful site covering most aspects of climate change is: climatekids.nasa.gov/
For a book in pdf form suitable for children visit:
www.unicef.org/zimbabwe/reports/child-friendly-climate-change-handbook
Two sites with New Zealand information and examples are:
www.mfe.govt.nz/climate-change
niwa.co.nz/education-and-training/schools/students/climate-change

1 Beginnings

bubbles	kids.kiddle.co/Egg_(biology)
	kids.kiddle.co/Seed
	kids.kiddle.co/Embryo
climate	climatekids.nasa.gov/weather-climate/
	climatekids.nasa.gov/climate-change-meaning/
climate survivor	www.doc.govt.nz/nature/native-animals/reptiles-and-frogs/tuatara/
	climatekids.nasa.gov/climate-change-evidence/
variation	eschooltoday.com/science/genetics/what-is-genetic-variation.html

2 Science of Climate Change

living things	www.kidsworldfun.com/learn-science/living-and-non-living-things.php
cells	kids.kiddle.co/Cell
atoms	kids.kiddle.co/Atom
	www.chem4kids.com/files/atom_intro.html
photosynthesis	photosynthesiseducation.com/photosynthesis-for-kids/
	kids.kiddle.co/Photosynthesis
light	www.ducksters.com/science/light.php
	www.explainthatstuff.com/light.html
	www.sciencekids.co.nz/light.html
greenhouse effect	kids.niehs.nih.gov/topics/natural-world/greenhouse-effect/index.htm
	climatekids.nasa.gov/greenhouse-effect/
	archive.epa.gov/climatechange/kids/basics/today/greenhouse-effect.html
	envis.tropmet.res.in/kidscorner/greenhouse.htm

3 Causes and Cures

fossil fuels
www.kidcyber.com.au/fossil-fuels
www.factsjustforkids.com/technology-facts/fossil-fuel-facts-for-kids.html
climatekids.nasa.gov/carbon/
climatekids.nasa.gov/fossil-fuels-coal/
climatekids.nasa.gov/air-pollution/

carbon footprint
www.parkcitygreen.org/Calculators/Kids-Calculator.aspx
calc.zerofootprint.net/
kids.lovetoknow.com/kids-activities/carbon-footprint-calculator-kids

methane
kids.kiddle.co/Methane
climatekids.nasa.gov/greenhouse-cards/
kidshealth.org/en/kids/fart.html
archive.epa.gov/climatechange/kids/solutions/technologies/methane.html

zero carbon
climatekids.nasa.gov/offset/

renewable energy
www.alliantenergykids.com/RenewableEnergy/RenewableEnergyHome
kids.kiddle.co/Renewable_resource
www.generationgenius.com/renewable-and-nonrenewable-energy-for-kids/

hydrogen fuel cell
kids.kiddle.co/Hydrogen_car
www.toyota.co.jp/en/kids/eco/fchv.html
www.eia.gov/kids/energy-sources/hydrogen/

4 Consequences

extreme weather
www.weatherwizkids.com/weather-safety.htm
www.oddizzi.com/teachers/explore-the-world/weather/extremes/
www.sciencekids.co.nz/videos/weather.html
climatekids.nasa.gov/10-things-glaciers/
climatekids.nasa.gov/sea-level/

ocean acidification
archive.epa.gov/climatechange/kids/impacts/signs/acidity.html
kids.kiddle.co/Coral_bleaching
climatekids.nasa.gov/acid-ocean/
climatekids.nasa.gov/ocean/
climatekids.nasa.gov/coral-bleaching/

mass extinctions
kids.earth.org/protecting-wildlife/what-is-the-sixth-mass-extinction/
kids.kiddle.co/Extinction

Index

A

acidification *42, 43*
agriculture *32*
algae *28, 33*
animal *28, 30, 31, 32, 40, 42, 43*
    animal - definition 16
Anthropocene extinction *44*
anthropogenic *44*
Arrhenius, Svante *44*
asteroid *14, 44*
atmosphere *7, 14, 21, 22, 26, 27, 28, 29, 30, 31, 32, 35, 37, 38, 42, 44, 46*
atom *16, 19, 21*

B

battery
    method 37
biomass *36, 38*
Bubble Earth *7, 9, 21, 28, 29, 30, 35, 41, 44, 47*
building blocks *16*

C

calcium carbonate *42, 43*
carbon *19, 20, 21*
carbon dioxide *9, 19, 20*
carbon footprint *31*
carbonic acid *42*
cells see *living blocks*
climate *9, 10, 13, 14, 18, 26, 39*
    New Zealand 11, 12
climate change *6, 7, 9, 10, 11, 12, 14, 15, 16, 18, 22, 42, 43, 44*
coal *22*, see also *fossil fuels*
coral *43*

D

DNA *12–56*

E

egg *8–56*
electric
    plane 38
    vehicle 37
electricity *33*
eructation and flatulence *32*
extinction
    mass extinction 44

F

fossil fuel *22*
fuel cell *37*

G

generation period *13*
glacier *39*
glucose *19*
greenhouse effect *26*
greenhouse gas *27*

H

H2O *38*
harpoon weed *33*
hydrogen *18*

I

Ice Age *11*
ice cap *41*
infrared *23*

K

kerosene *38*
Kiribati *41*

L

light energy *50*
limestone *42*
living blocks *16*

M

methane *32*
molecule *18, 20, 21, 28*

N

New Zealand *11*
    carbon footprint 32
    climate change 12
    hydroelectricity 36
    recycling 47
nitrous oxide *18, 27*

O

ocean *8, 36, 39*
    ocean acidification 42
oil see *fossil fuels*
oxygen *8, 18, 19, 20, 25, 28, 29, 30, 37*

P

petroleum oil *37*
    petrol 28
photon *22, 23, 24, 25, 26, 36, 41*
photosynthesis *19, 20, 21, 22, 25, 28, 29, 30, 32, 34*
plants *16, 28*
    energy 19
    fossilised 22
    photosynthesis 29
Priestley, Joseph *15*

R

renewable energy *35*
respiration *20, 21*

S

sea level *11, 41*
sex (determination) *12*
solar farms *36*

T

thermal *23, 24, 25, 26, 27, 32, 35*
tuatara *10, 11, 12, 13, 14*
Tuvalu *41*

U

ultraviolet *23, 24*

V

variation *13*

W

water *8, 11, 18, 19, 38*
    rainwater 36
    vapour 27
weather *9, 39*
    extreme events 40